NASA
Reference
Publication
1228

1990

Fastener Design Manual

Richard T. Barrett
Lewis Research Center
Cleveland, Ohio

National Aeronautics and
Space Administration

Office of Management

Scientific and Technical
Information Division

ERRATA

NASA Reference Publication 1228

Fastener Design Manual

Richard T. Barrett

March 1990

The manual describes various platings that may be used for corrosion control including cadmium and zinc plating. It does not mention outgassing problems caused by the relatively high vapor pressure of these metals. The fastener manual was intended primarily for aeronautical applications, where outgassing is typically not a concern.

Issued June 17, 2008

Contents

	Page
Summary	1
Introduction	1

General Design Information

Fastener Materials	1
Platings and Coatings	1
Thread Lubricants	4
Corrosion	5
Locking Methods	6
Washers	9
Inserts	10
Threads	12
Fatigue-Resistant Bolts	13
Fastener Torque	15
Design Criteria	17

Rivets and Lockbolts

Rivets	26
Lockbolts	30
General Guidelines for Selecting Rivets and Lockbolts	34

References	35

Appendixes

A—Bolthead Marking and Design Data	36
B—Bolt Ultimate Shear and Tensile Strengths	90
C—Blind Rivet Requirements	94

Summary

This manual was written for design engineers to enable them to choose appropriate fasteners for their designs. Subject matter includes fastener material selection, platings, lubricants, corrosion, locking methods, washers, inserts, thread types and classes, fatigue loading, and fastener torque. A section on design criteria covers the derivation of torque formulas, loads on a fastener group, combining simultaneous shear and tension loads, pullout load for tapped holes, grip length, head styles, and fastener strengths. The second half of this manual presents general guidelines and selection criteria for rivets and lockbolts.

Introduction

To the casual observer the selection of bolts, nuts, and rivets for a design should be a simple task. In reality it is a difficult task, requiring careful consideration of temperature, corrosion, vibration, fatigue, initial preload, and many other factors.

The intent of this manual is to present enough data on bolt and rivet materials, finishes, torques, and thread lubricants to enable a designer to make a sensible selection for a particular design. Locknuts, washers, locking methods, inserts, rivets, and tapped holes are also covered.

General Design Information

Fastener Materials

Bolts can be made from many materials, but most bolts are made of carbon steel, alloy steel, or stainless steel. Stainless steels include both iron- and nickel-based chromium alloys. Titanium and aluminum bolts have limited usage, primarily in the aerospace industry.

Carbon steel is the cheapest and most common bolt material. Most hardware stores sell carbon steel bolts, which are usually zinc plated to resist corrosion. The typical ultimate strength of this bolt material is 55 ksi.

An alloy steel is a high-strength carbon steel that can be heat treated up to 300 ksi. However, it is not corrosion resistant and must therefore have some type of coating to protect it from corrosion. Aerospace alloy steel fasteners are usually cadmium plated for corrosion protection.

Bolts of stainless steel (CRES) are available in a variety of alloys with ultimate strengths from 70 to 220 ksi. The major advantage of using CRES is that it normally requires no protective coating and has a wider service temperature range than plain carbon or alloy steels.

A partial listing of bolt materials is given in table I. The following precautions are to be noted:

(1) The bolt plating material is usually the limiting factor on maximum service temperature.

(2) Carbon steel and alloy steel are unsatisfactory (become brittle) at temperatures below −65 °F.

(3) Hydrogen embrittlement is a problem with most common methods of plating, unless special procedures are used. (This subject is covered more fully in the corrosion section.)

(4) Series 400 CRES contains only 12 percent chromium and thus will corrode in some environments.

(5) The contact of dissimilar materials can create galvanic corrosion, which can become a major problem. (Galvanic corrosion is covered in a subsequent section of this manual.)

Platings and Coatings

Most plating processes are electrolytic and generate hydrogen. Thus, most plating processes require baking after plating at a temperature well below the decomposition temperature of the plating material to prevent hydrogen embrittlement. However, heating the plating to its decomposition temperature can generate free hydrogen again. Thus, exceeding the safe operating temperature of the plating can cause premature fastener failure due to hydrogen embrittlement as well as loss of corrosion protection. (A summary of platings and coatings is given in table II.)

Cadmium Plating

The most common aerospace fastener plating material is cadmium. Plating is done by electrodeposition and is easy to accomplish. However, cadmium-plated parts must be baked at 375 °F for 23 hours, within 2 hours after plating, to prevent hydrogen embrittlement. Since cadmium melts at 600 °F, its useful service temperature limit is 450 °F.

TABLE I.—SUMMARY OF FASTENER MATERIALS

Material	Surface treatment	Useful design temperature limit, °F	Ultimate tensile strength at room temperature, ksi	Comments
Carbon steel	Zinc plate	−65 to 250	55 and up	-------------------
Alloy steels	Cadmium plate, nickel plate, zinc plate, or chromium plate	−65 to limiting temperature of plating	Up to 300	Some can be used at 900 °F
A-286 stainless	Passivated per MIL-S-5002	−423 to 1200	Up to 220	-------------------
17-4PH stainless	None	−300 to 600	Up to 220	-------------------
17-7PH stainless	Passivated	−200 to 600	Up to 220	-------------------
300 series stainless	Furnace oxidized	−423 to 800	70 to 140	Oxidation reduces galling
410, 416, and 430 stainless	Passivated	−250 to 1200	Up to 180	47 ksi at 1200 °F; will corrode slightly
U-212 stainless	Cleaned and passivated per MIL-S-5002	1200	185	140 ksi at 1200 °F
Inconel 718 stainless	Passivated per QQ-P-35 or cadmium plated	−423 to 900 or cadmium plate limit	Up to 220	-------------------
Inconel X-750 stainless	None	−320 to 1200	Up to 180	136 ksi at 1200 °F
Waspalloy stainless	None	−423 to 1600	150	-------------------
Titanium	None	−350 to 500	Up to 160	-------------------

Zinc Plating

Zinc is also a common type of plating. The hot-dip method of zinc plating is known commercially as galvanizing. Zinc can also be electrodeposited. Because zinc plating has a dull finish, it is less pleasing in appearance than cadmium. However, zinc is a sacrificial material. It will migrate to uncoated areas that have had their plating scratched off, thus continuing to provide corrosion resistance. Zinc may also be applied cold as a zinc-rich paint. Zinc melts at 785 °F but has a useful service temperature limit of 250 °F. (Its corrosion-inhibiting qualities degrade above 140 °F.)

Phosphate Coatings

Steel or iron is phosphate coated by treating the material surface with a diluted solution of phosphoric acid, usually by submerging the part in a proprietary bath. The chemical reaction forms a mildly protective layer of crystalline phosphate. The three principal types of phosphate coatings are zinc, iron, and manganese. Phosphate-coated parts can be readily painted, or they can be dipped in oil or wax to improve their corrosion resistance. Fasteners are usually phosphate with either zinc or manganese. Hydrogen embrittlement seldom is present in phosphated parts. Phosphate coatings start deteriorating at 225 °F (for heavy zinc) to 400 °F (for iron phosphate).

Nickel Plating

Nickel plating, with or without a copper strike (thin plating) is one of the oldest methods of preventing corrosion and improving the appearance of steel and brass. Nickel plating will tarnish unless it is followed by chromium plating. Nickel plating is a more expensive process than cadmium or zinc plating and also must be baked the same as cadmium after plating to prevent hydrogen embrittlement. Nickel plating good to an operating temperature of 1100 °F, but is still not frequently used for plating fasteners because of its cost.

TABLE II.—SUMMARY OF PLATINGS AND COATINGS

Type of coating	Useful design temperature limit, °F	Remarks
Cadmium	450	Most common for aerospace fasteners
Zinc	140 to 250	Self-healing and cheaper than cadmium
Phosphates: Manganese Zinc Iron	225 225 to 375 400	Mildly corrosion resistant but main use is for surface treatment prior to painting. Another use is with oil or wax for deterring corrosion.
Chromium	800 to 1200	Too expensive for most applications other than decorative
Silver	1600	Most expensive coating
Black oxide (and oil)	[a]300	Ineffective in corrosion prevention
Preoxidation (CRES) fasteners only	1200	Prevents freeze-up of CRES threads due to oxidation after installation
Nickel	1100	More expensive than cadmium or zinc
SermaGard and Sermatel W	450 to 1000	Dispersed aluminum particles with chromates in a water-based ceramic base coat
Stalgard	475	Proprietary organic and/or organic-inorganic compound used for corrosion resistance and lubrication (in some cases)
Diffused nickel-cadmium	900	Expensive and requires close control to avoid hydrogen damage

[a]Oil boiling point.

Ion-Vapor-Deposited Aluminum Plating

Ion-vapor-deposited aluminum plating was developed by McDonnell-Douglas for coating aircraft parts. It has some advantages over cadmium plating:

(1) It creates no hydrogen embrittlement.
(2) It insulates against galvanic corrosion of dissimilar materials.
(3) The coating is acceptable up to 925 °F.
(4) It can also be used for coating titanium and aluminums.
(5) No toxic byproducts are formed by the process.
It also has some disadvantages:
(1) Because the process must be done in a specially designed vacuum chamber, it is quite expensive.
(2) Cadmium will outperform ion-vapor-deposited aluminum in a salt-spray test.

Chromium Plating

Chromium plating is commonly used for automotive and appliance decorative applications, but it is not common for fasteners. Chromium-plated fasteners cost approximately as much as stainless steel fasteners. Good chromium plating requires both copper and nickel plating prior to chromium plating. Chromium plating also has hydrogen embrittlement problems. However, it is acceptable for maximum operating temperatures of 800 to 1200 °F.

Sermatel W and SermaGard

Sermatel W and SermaGard are proprietary coatings[1] consisting of aluminum particles in an inorganic binder with chromates added to inhibit corrosion. The coating material is covered by AMS3126A, and the procedure for applying it by AMS2506. The coating is sprayed or dipped on the part and cured at 650 °F. (SPS Technologies[2] has tested Sermatel W-coated fasteners at 900 °F without degradation.) This coating process prevents both hydrogen embrittlement and stress corrosion, since the fastener is completely coated. Sermatel is about as effective as cadmium plating in resisting corrosion but costs about 15 percent more than cadmium. Fasteners are not presently available "off the shelf" with Sermatel W or SermaGard coating, but the company will do small orders for fasteners or mechanical parts. These coatings will take up to 15 disassemblies in a threaded area without serious coating degradation.

Stalgard

Stalgard is a proprietary coating[3] process consisting of organic coatings, inorganic-organic coatings, or both for corrosion resistance. According to Stalgard test data their coatings are superior to either cadmium or zinc plating in salt-spray and weathering tests. Stalgard coatings also provide galvanic corrosion protection. However, the maximum operating temperature of these organic coatings is 475 °F.

Diffused Nickel-Cadmium Plating

This process was developed by the aerospace companies for a higher temperature cadmium coating. A 0.0004-in.-thick nickel coating is plated on the substrate, followed by a 0.0002-in.-thick. cadmium plate (per AMS2416). The part is then baked for 1 hour at 645 °F. The resulting coating can withstand 1000 °F. However, the nickel plate must completely cover the part at all times to avoid cadmium damage to the part. This process is expensive and requires close control.

[1]Sermatech International, Inc., Limerick, Pennsylvania.
[2]Jenkintown, Pennsylvania.
[3]Elco Industries, Rockford, Illinois.

Silver Plating

Silver plating is cost prohibitive for most fastener applications. The big exception is in the aerospace industry, where silver-plated nuts are used on stainless steel bolts. The silver serves both as a corrosion deterrent and a dry lubricant. Silver plating can be used to 1600 °F, and thus it is a good high-temperature lubricant. Since silver tarnishes from normal atmospheric exposure, the silver-plated nuts are commonly coated with clear wax to prevent tarnishing. Wax is a good room-temperature lubricant. Therefore, the normal "dry torque" values of the torque tables should be reduced by 50 percent to allow for this lubricant.

Passivation and Preoxidation

Stainless steel fasteners will create galvanic corrosion or oxidation in a joint unless they are passivated or preoxidized prior to assembly (ref. 1). Passivation is the formation of a protective oxide coating on the steel by treating it briefly with an acid. The oxide coating is almost inert. Preoxidization is the formation of an oxide coating by exposing the fasteners to approximately 1300 °F temperature in an air furnace. The surface formed is inert enough to prevent galling due to galvanic corrosion.

Black Oxide Coating

Black oxide coating, combined with an oil film, does little more than enhance the appearance of carbon steel fasteners. The oil film is the only part of the coating that prevents corrosion.

Thread Lubricants

Although there are many thread lubricants from which to choose, only a few common ones are covered here. The most common are oil, grease or wax, graphite, and molybdenum disulfide. There are also several proprietary lubricants such as Never-Seez and Synergistic Coatings. Some thread-locking compounds such as Loctite can also be used as lubricants for a bolted assembly, particularly the compounds that allow the bolts to be removed. A summary of thread lubricants is given in table III.

Oil and Grease

Although oil and grease are the most common types of thread lubricants, they are limited to an operating temperature not much greater than 250 °F. (Above this temperature the oil or grease will melt or boil off.) In addition, oil cannot be used in a vacuum environment. However, oil and grease are good for both lubrication and corrosion prevention as long as these precautions are observed.

TABLE III.—SUMMARY OF THREAD LUBRICANTS

Type of lubricant	Useful design temperature limit, °F	Remarks
Oil or grease	250	Most common; cannot be used in vacuum
Graphite	[a]212 to 250	Cannot be used in vacuum
Molybdenum disulfide	750	Can be used in vacuum
Synergistic Coatings	500	Can be used in vacuum
Neverseez	2200	Because oil boils off, must be applied after each high-temperature application
Silver Goop	1500	Do not use on aluminum or magnesium parts; extremely expensive
Thread-locking compounds	275	"Removable fastener" compounds only

[a]Carrier boiloff temperature.

Graphite

"Dry" graphite is really not dry. It is fine carbon powder that needs moisture (usually oil or water) to become a lubricant. Therefore, its maximum operating temperature is limited to the boiling point of the oil or water. It also cannot be used in a vacuum environment without losing its moisture. Because dry graphite is an abrasive, its use is detrimental to the bolted joint if the preceding limitations are exceeded.

Molybdenum Disulfide

Molybdenum disulfide is one of the most popular dry lubricants. It can be used in a vacuum environment but turns to molybdenum trisulfide at approximately 750 °F. Molybdenum trisulfide is an abrasive rather than a lubricant.

Synergistic Coatings

These proprietary coatings[4] are a type of fluorocarbon injected and baked into a porous metal-matrix coating to give both corrosion prevention and lubrication. However, the maximum operating temperature given in their sales literature is 500 °F. Synergistic Coatings will also operate in a vacuum environment.

Neverseez

This proprietary compound[5] is a petroleum-base lubricant and anticorrodent that is satisfactory as a one-time lubricant

---

[4]General Magnaplate Corporation, Ventura, California.
[5]Bostic Emhart, Broadview, Illinois.

up to 2200 °F, according to the manufacturer. The oil boils off, but the compound leaves nongalling oxides of nickel, copper, and zinc between the threads. This allows the fastener to be removed, but a new application is required each time the fastener is installed. NASA Lewis personnel tested this compound and found it to be satisfactory.

Silver Goop

Silver Goop is a proprietary compound[6] containing 20 to 30 percent silver. Silver Goop can be used to 1500 °F, but it is not to be used on aluminum or magnesium. It is extremely expensive because of its silver content.

Thread-Locking Compounds

Some of the removable thread-locking compounds (such as Loctite) also serve as antigalling and lubricating substances. However, they are epoxies, which have a maximum operating temperature of approximately 275 °F.

Corrosion

Galvanic Corrosion

Galvanic corrosion is set up when two dissimilar metals are in the presence of an electrolyte, such as moisture. A galvanic cell is created and the most active (anode) of the two materials is eroded and deposited on the least active (cathode). Note that the farther apart two materials are in the following list, the greater the galvanic action between them.

According to reference 2 the galvanic ranking of some common engineering materials is as follows:

(1) Magnesium (most active)
(2) Magnesium alloys
(3) Zinc
(4) Aluminum 5056
(5) Aluminum 5052
(6) Aluminum 1100
(7) Cadmium
(8) Aluminum 2024
(9) Aluminum 7075
(10) Mild steel
(11) Cast iron
(12) Ni-Resist
(13) Type 410 stainless (active)
(14) Type 304 stainless (active)
(15) Type 316 stainless (active)
(16) Lead
(17) Tin
(18) Muntz Metal
(19) Nickel (active)

(20) Inconel (active)
(21) Yellow brass
(22) Admiralty brass
(23) Aluminum brass
(24) Red brass
(25) Copper
(26) Silicon bronze
(27) 70–30 Copper-nickel
(28) Nickel (passive)
(29) Inconel (passive)
(30) Titanium
(31) Monel
(32) Type 304 stainless (passive)
(33) Type 316 stainless (passive)
(34) Silver
(35) Graphite
(36) Gold (least active)

Note the difference between active and passive 304 and 316 stainless steels. The difference here is that passivation of stainless steels is done either by oxidizing in an air furnace or treating the surface with an acid to cause an oxide to form. This oxide surface is quite inert in both cases and deters galvanic activity.

Because the anode is eroded in a galvanic cell, it should be the larger mass in the cell. Therefore, it is poor design practice to use carbon steel fasteners in a stainless steel or copper assembly. Stainless steel fasteners can be used in carbon steel assemblies, since the carbon steel mass is the anode.

Magnesium is frequently used in lightweight designs because of its high strength to weight ratio. However, it must be totally insulated from fasteners by an inert coating such as zinc chromate primer to prevent extreme galvanic corrosion. Cadmium- or zinc-plated fasteners are closest to magnesium in the galvanic series and would be the most compatible if the insulation coating were damaged.

Stress Corrosion

Stress corrosion occurs when a tensile-stressed part is placed in a corrosive environment. An otherwise ductile part will fail at a stress much lower than its yield strength because of surface imperfections (usually pits or cracks) created by the corrosive environment. In general, the higher the heat-treating temperature of the material (and the lower the ductility), the more susceptible it is to stress corrosion cracking.

The fastener material manufacturers have been forced to develop alloys that are less sensitive to stress corrosion. Of the stainless steels, A286 is the best fastener material for aerospace usage. It is not susceptible to stress corrosion but usually is produced only up to 160-ksi strength (220-ksi A286 fasteners are available on special order). The higher strength stainless steel fasteners (180 to 220 ksi) are usually made of 17–7PH or 17–4PH, which are stress corrosion susceptible. Fasteners made of superalloys such as Inconel 718 or MP35N are available if cost and schedule are not restricted.

6Swagelok Company, Solon, Ohio.

An alternative is to use a high-strength carbon steel (such as H–11 tool steel with an ultimate tensile strength of 300 ksi) and provide corrosion protection. However, it is preferable to use more fasteners of the ordinary variety and strength, if possible, than to use a few high-strength fasteners. High-strength fasteners (greater than 180 ksi) bring on problems such as brittleness, critical flaws, forged heads, cold rolling of threads, and the necessity for stringent quality control procedures. Quality control procedures such as x-ray, dye penetrant, magnetic particle, thread radius, and head radius inspections are commonly used for high-strength fasteners.

Hydrogen Embrittlement

Hydrogen embrittlement occurs whenever there is free hydrogen in close association with the metal. Since most plating processes are the electrolytic bath type, free hydrogen is present. There are three types of hydrogen-metal problems:

(1) Hydrogen chemical reaction: Hydrogen reacts with the carbon in steel to form methane gas, which can lead to crack development and strength reduction. Hydrogen can also react with alloying elements such as titanium, niobium, or tantalum to form hydrides. Because the hydrides are not as strong as the parent alloy, they reduce the overall strength of the part.

(2) Internal hydrogen embrittlement: Hydrogen can remain in solution interstitially (between lattices in the grain structure) and can cause delayed failures after proof testing. There is no external indication that the hydrogen is present.

(3) Hydrogen environment embrittlement: This problem is only present in a high-pressure hydrogen environment such as a hydrogen storage tank. Unless a fastener was under stress inside such a pressure vessel, this condition would not be present.

Most plating specifications now state that a plated carbon steel fastener "shall be baked for not less than 23 hours at 375 ± 25 °F within 2 hours after plating to provide hydrogen embrittlement relief" (per MIL–N–25027D). In the past the plating specifications required baking at 375 ± 25 °F for only 3 hours within 4 hours after plating. This treatment was found to be inadequate, and most plating specifications were revised in 1981–82 to reflect the longer baking time. Hydrogen embrittlement problems also increase as the fastener strength increases.

Cadmium Embrittlement

Although hydrogen embrittlement failure of materials is well documented (ref. 3), the effects of cadmium embrittlement are not. In general, hydrogen embrittlement failure of cadmium-plated parts can start as low as 325 °F, but cadmium embrittlement can start around 400 °F. Since both elements are normally present in elevated-temperature failure of cadmium-plated parts, the combined effect of the two can be disastrous. However, the individual effect of each is indeterminate.

Locking Methods

Tapped Holes

In a tapped hole the locking technique is normally on the fastener. One notable exception is the Spiralock[7] tap shown in figure 1. The Spiralock thread form has a 30° wedge ramp at its root. Under clamp load the crests of the male threads are wedged tightly against the ramp. This makes lateral movement, which causes loosening under vibration, nearly impossible. Independent tests by some of the aerospace companies have indicated that this type of thread is satisfactory for moderate resistance to vibration. The bolt can have a standard thread, since the tapped hole does all the locking.

Locknuts

There are various types of locking elements, with the common principle being to bind (or wedge) the nut thread to the bolt threads. Some of the more common locknuts are covered here.

Split beam.—The split-beam locknut (fig. 2) has slots in the top, and the thread diameter is undersized in the slotted portion. The nut spins freely until the bolt threads get to the slotted area. The split "beam" segments are deflected outward by the bolt, and a friction load results from binding of the mating threads.

Figure 1.—Spiralock thread.

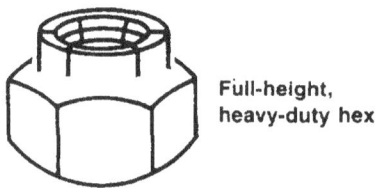

Figure 2.—Split-beam locknut.

---

[7]Distributed by Detroit Tap & Tool Company, Detroit, Michigan, through license from H.D. Holmes.

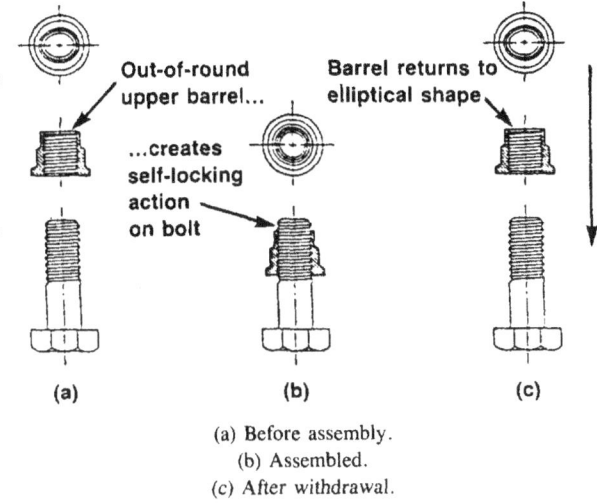

(a) Before assembly.
(b) Assembled.
(c) After withdrawal.

Figure 3.—Deformed-thread locknut.

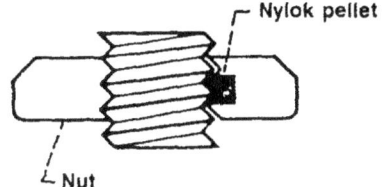

Figure 4.—Nylok pellet locknut.

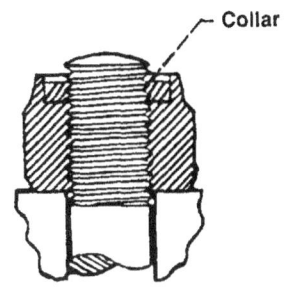

Figure 5.—Locking collar.

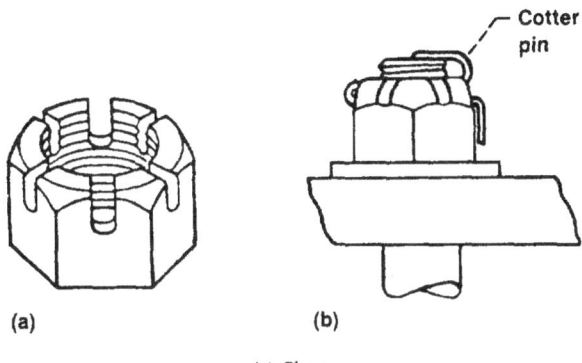

(a) Slots.
(b) Cotter pin locking.

Figure 6.—Castellated nut.

Deformed thread.—The deformed-thread locknut (fig. 3) is a common locknut, particularly in the aerospace industry. Its advantages are as follows:

(1) The nut can be formed in one operation.

(2) The temperature range is limited only by the parent metal, its plating, or both.

(3) The nut can be reused approximately 10 times before it has to be discarded for loss of locking capability.

Nylok pellet.—The Nylok[8] pellet (of nylon) is usually installed in the nut threads as shown in figure 4. A pellet or patch projects from the threads. When mating threads engage, compression creates a counterforce that results in locking contact. The main drawback of this pellet is that its maximum operating temperature is approximately 250 °F. The nylon pellet will also be damaged quickly by reassembly.

Locking collar and seal.—A fiber or nylon washer is mounted in the top of the nut as shown in figure 5. The collar has an interference fit such that it binds on the bolt threads. It also provides some sealing action from gas and moisture leakage. Once again the limiting feature of this nut is the approximate 250 °F temperature limit of the locking collar.

A cost-saving method sometimes used instead of a collar or nylon pellet is to bond a nylon patch on the threads of either the nut or the bolt to get some locking action. This method is also used on short thread lengths, where a drilled hole for a locking pellet could cause severe stress concentration.

Castellated nut.—The castellated nut normally has six slots as shown in figure 6(a). The bolt has a single hole through its threaded end. The nut is torqued to its desired torque value. It is then rotated forward or backward (depending on the user's

preference) to the nearest slot that aligns with the drilled hole in the bolt. A cotter pin is then installed to lock the nut in place as shown in figure 6(b). This nut works extremely well for low-torque applications such as holding a wheel bearing in place.

Jam nuts.—These nuts are normally "jammed" together as shown in figure 7, although the "experts" cannot agree on which nut should be on the bottom. However, this type of assembly is too unpredictable to be reliable. If the inner nut is torqued tighter than the outer nut, the inner nut will yield before the outer nut can pick up its full load. On the other hand, if the outer nut is tightened more than the inner nut, the inner nut unloads. Then the outer nut will yield before the inner nut can pick up its full load. It would be rare to get the correct amount of torque on each nut. A locknut is a much more practical choice than a regular nut and a jam nut. However, a jam nut can be used on a turnbuckle, where it does not carry any of the tension load.

---

[8]Nylok Fastener Corporation, Rochester, Michigan.

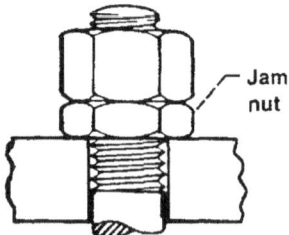

Figure 7.—Jam nut.

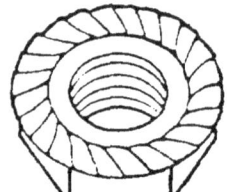

Figure 8.—Durlock nut.

Serrated-face nut (or bolthead).—The serrated face of this nut (shown in fig. 8) digs into the bearing surface during final tightening. This means that it cannot be used with a washer or on surfaces where scratches or corrosion could be a problem.

According to SPS Technologies, their serrated-face bolts (Durlock 180) require 110 percent of tightening torque to loosen them. Their tests on these bolts have shown them to have excellent vibration resistance.

Lockwiring.—Although lockwiring is a laborious method of preventing bolt or nut rotation, it is still used in critical applications, particularly in the aerospace field. The nuts usually have drilled corners, and the bolts either have throughholes in the head or drilled corners to thread the lockwire through. A typical bolthead lockwiring assembly is shown in figure 9(a), and a typical nut lockwiring assembly is shown in figure 9(b).

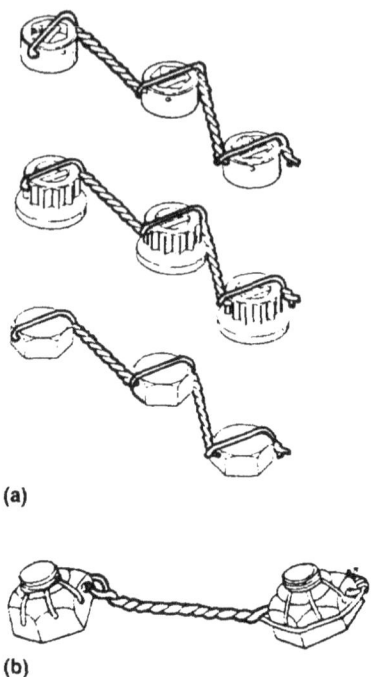

(a)

(b)

(a) Multiple fastener application (double-twist method, single hole).
(b) Castellated nuts on undrilled studs (double-twist method).

Figure 9.—Lockwiring.

Direct interfering thread.—A direct interfering thread has an oversized root diameter that gives a slight interference fit between the mating threads. It is commonly used on threaded studs for semipermanent installations, rather than on bolts and nuts, since the interference fit does damage the threads.

Tapered thread.—The tapered thread is a variation of the direct interfering thread, but the difference is that the minor diameter is tapered to interfere on the last three or four threads of a nut or bolt as shown in figure 10.

Nutplates.—A nutplate (fig. 11) is normally used as a blind nut. They can be fixed or floating. In addition, they can have

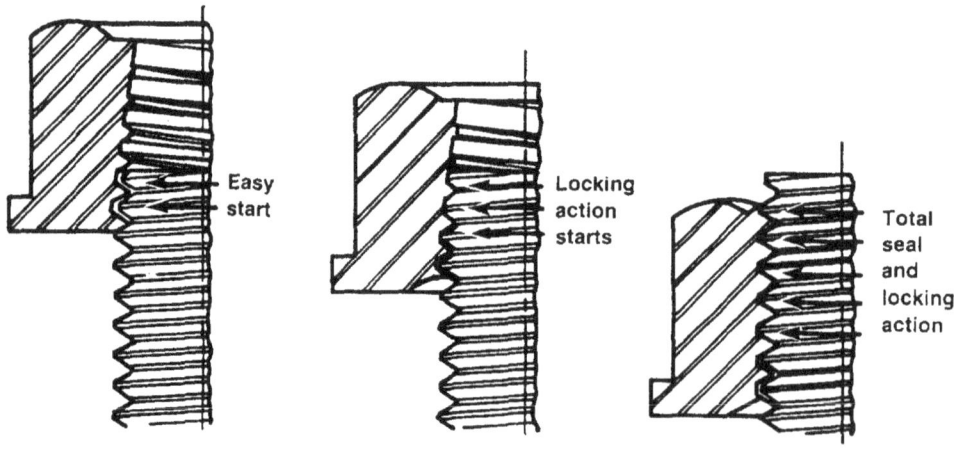

Figure 10.—Tapered thread.

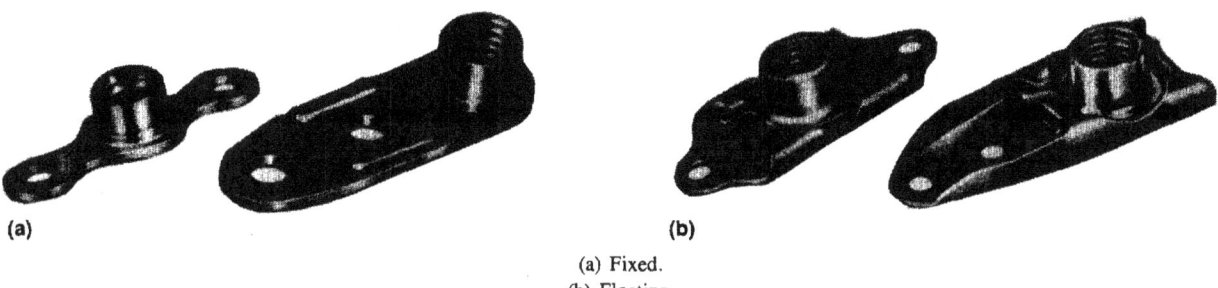

(a) Fixed.
(b) Floating.

Figure 11.—Nutplate.

most of the locking and sealing features of a regular nut. Nutplates are usually used on materials too thin to tap. They are used primarily by the aerospace companies, since their installation is expensive. At least three drilled holes and two rivets are required for each nutplate installation.

Locking Adhesives

Many manufacturers make locking adhesives (or epoxies) for locking threads. Most major manufacturers make several grades of locking adhesive, so that the frequency of disassembly can be matched to the locking capability of the adhesive. For example, Loctite 242 is for removable fasteners, and Loctite 271[9] is for tamperproof fasteners. Other manufacturers such as Bostik, ND Industries, Nylock, 3M, and Permaloc make similar products.

Most of these adhesives work in one of two ways. They are either a single mixture that hardens when it becomes a thin layer in the absence of air or an epoxy in two layers that does not harden until it is mixed and compressed between the mating threads. Note that the two-layer adhesives are usually put on the fastener as a "ribbon" or ring by the manufacturer. These ribbons or rings do have some shelf life, as long as they are not inadvertently mixed or damaged.

These adhesives are usually effective as thread sealers as well. However, *none of them will take high temperatures.* The best adhesives will function at 450 °F; the worst ones will function at only 200 °F.

Washers

Belleville Washers

Belleville washers (fig. 12) are conical washers used more for maintaining a uniform tension load on a bolt than for locking. If they are not completely flattened out, they serve as a spring in the bolt joint. However, unless they have serrations on their surfaces, they have no significant locking capability. Of course, the serrations will damage the mating surfaces under them. These washers can be stacked in

combinations as shown in figure 13 to either increase the total spring length (figs. 13(a) and (c)) or increase the spring constant (fig. 13(b)).

Lockwashers

The typical helical spring washer shown in figure 14 is made of slightly trapezoidal wire formed into a helix of one coil so that the free height is approximately twice the thickness of the washer cross section. They are usually made of hardened carbon steel, but they are also available in aluminum, silicon, bronze, phosphor-bronze, stainless steel, and K-Monel.

The lockwasher serves as a spring while the bolt is being tightened. However, the washer is normally flat by the time the bolt is fully torqued. At this time it is equivalent to a solid flat washer, and its locking ability is nonexistent. In summary, a lockwasher of this type is useless for locking.

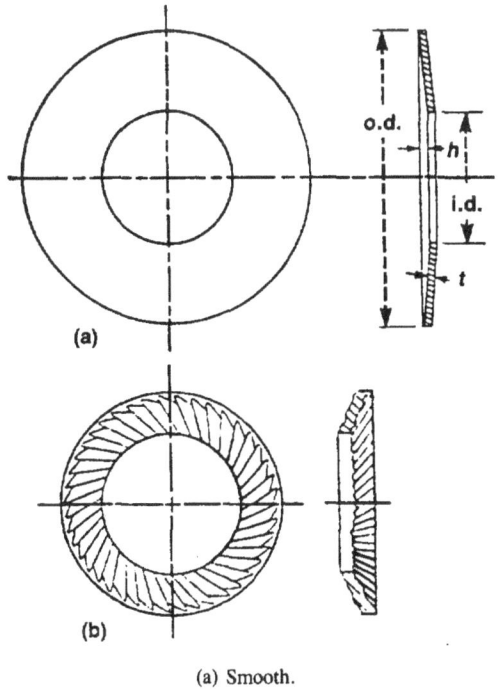

(a) Smooth.
(b) Serrated.

Figure 12.—Types of Belleville washers.

[9]*Loctite Corporation, Newington, Connecticut.*

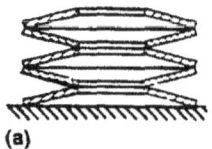

(a)

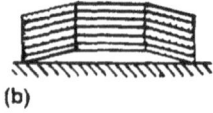

(b)

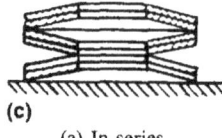

(c)

(a) In series.
(b) In parallel.
(c) In-parallel series.
Figure 13.—Combinations of Belleville washers.

Figure 14.—Helical spring washers.

Tooth (or Star) Lockwashers

Tooth lockwashers (fig. 15) are used with screws and nuts for some spring action but mostly for locking action. The teeth are formed in a twisted configuration with sharp edges. One edge bites into the bolthead (or nut) while the other edge bites into the mating surface. Although this washer does provide some locking action, it damages the mating surfaces. These scratches can cause crack formation in highly stressed fasteners, in mating parts, or both, as well as increased corrosion susceptibility.

Self-Aligning Washers

A self-aligning washer is used with a mating nut that has conical faces as shown in figure 16. Because there is both a weight penalty and a severe cost penalty for using this nut, it should be used only as a last resort. Maintaining parallel mating surfaces within acceptable limits (2° per SAE Handbook (ref. 4)) is normally the better alternative.

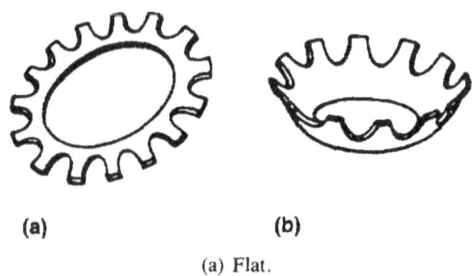

(a)   (b)

(a) Flat.
(b) Countersunk.
Figure 15.—Tooth lockwashers.

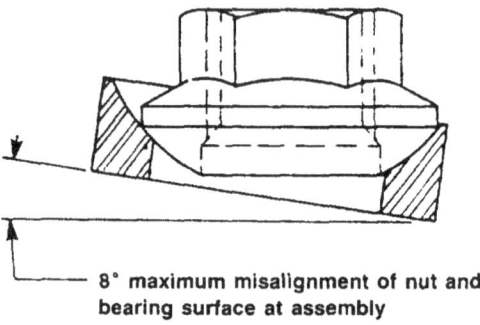

8° maximum misalignment of nut and bearing surface at assembly

Figure 16.—Self-aligning nut.

Inserts

An insert is a special type of device that is threaded on its inside diameter and locked with threads or protrusions on its outside diameter in a drilled, molded, or tapped hole. It is used to provide a strong, wear-resistant tapped hole in a soft material such as plastic and nonferrous materials, as well as to repair stripped threads in a tapped hole.

The aerospace industry uses inserts in tapped holes in soft materials in order to utilize small high-strength fasteners to save weight. The bigger external thread of the insert (nominally 1/8 in. bigger in diameter than the internal thread) gives, for example, a 10–32 bolt in an equivalent 5/16–18 nut.

In general, there are two types of inserts: those that are threaded externally, and those that are locked by some method other than threads (knurls, serrations, grooves, or interference fit). Within the threaded inserts there are three types: the wire thread, the self-tapping, and the solid bushing.

Threaded Inserts

Wire thread.—The wire thread type of insert (Heli-coil[10])

---

[10]Emhart Fastening Systems Group, Heli-Coil Division, Danbury, Connecticut.

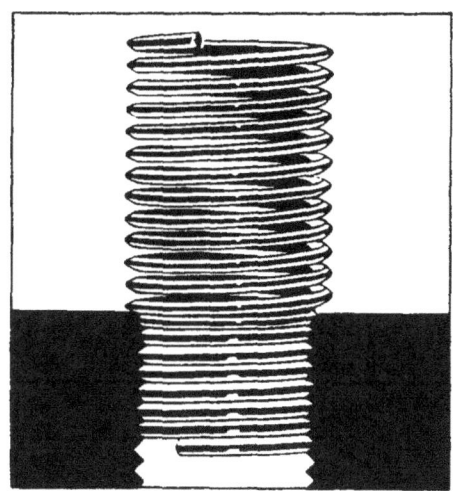

Figure 17.—Wire thread insert installation.

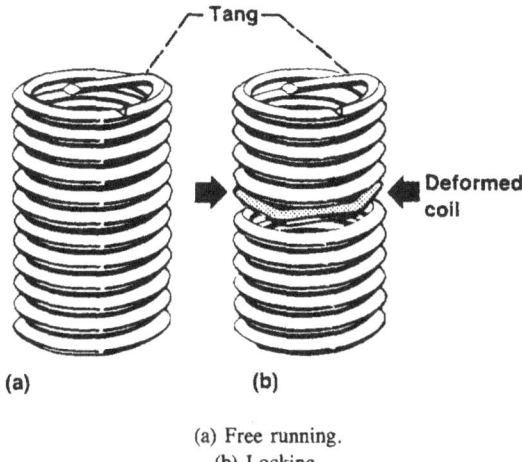

(a) Free running.
(b) Locking.

Figure 18.—Wire thread insert types.

is a precision coil of diamond-shaped CRES wire that forms both external and internal threads as shown in figure 17. The coil is made slightly oversize so that it will have an interference fit in the tapped hole. In addition, this insert is available with a deformed coil (fig. 18) for additional locking. The tang is broken off at the notch after installation.

The wire thread insert is the most popular type for repair of a tapped hole with stripped threads, since it requires the least amount of hole enlargement. However, the solid bushing insert is preferred if space permits.

Self-tapping.—Most of the self-tapping inserts are the solid bushing type made with a tapered external thread similar to a self-tapping screw (fig. 19). There are several different

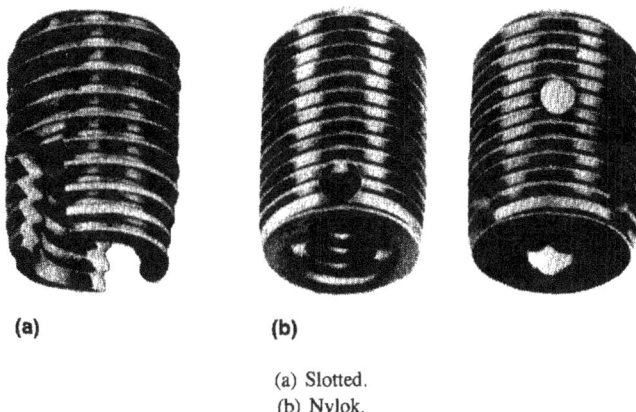

(a)   (b)

(a) Slotted.
(b) Nylok.

Figure 19.—Self-tapping inserts.

locking combinations, such as the Nylok plug (fig. 19(b)) or the thread-forming Speedsert[11] deformed thread (fig. 20). An additional advantage of the thread-forming insert is that it generates no cutting chips, since it does not cut the threads. However, it can only be used in softer materials.

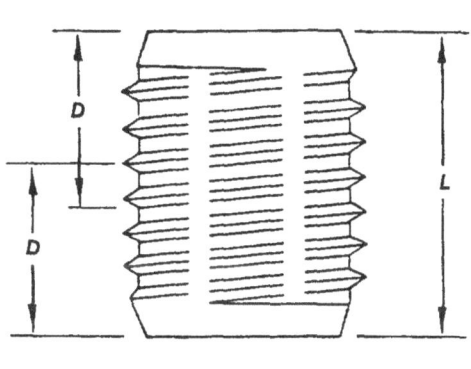

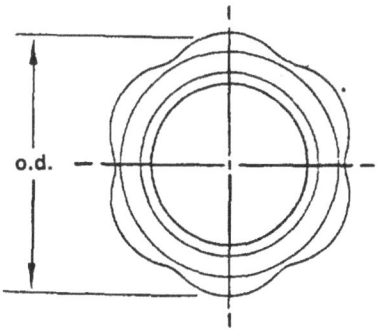

Figure 20.—Speedsert.

---

[11]Rexnord Specialty Fasteners Division, Torrance, California.

Solid bushing.—Solid bushing inserts have conventional threads both internally and externally. A popular type is the Keensert[11] shown in figure 21. The locking keys are driven in after the insert is in place. Another manufacturer uses a two-prong ring for locking. These inserts are also available with distorted external thread or Nylok plugs for locking.

Nonthreaded Inserts

Plastic expandable.—The most familiar of the nonthreaded inserts is the plastic expandable type shown in figure 22. This insert has barbs on the outside and longitudinal slits that allow it to expand outward as the threaded fastener is installed, pushing the barbs into the wall of the drilled hole. (See ref. 5.)

Molded in place.—This type of insert (fig. 23) is knurled or serrated to resist both pullout and rotation. It is commonly used with ceramics, rubber, and plastics, since it can develop higher resistance to both pullout and rotation in these materials than self-tapping or conventionally threaded inserts. (See ref. 5.)

Ultrasonic.—Ultrasonic inserts (fig. 24) have grooves in various directions to give them locking strength. They are installed in a prepared hole by pushing them in while they are being ultrasonically vibrated. The ultrasonic vibration melts the wall of the hole locally so that the insert grooves are "welded" in place. Since the area melted is small, these inserts do not have the holding power of those that are molded in place. Ultrasonic inserts are limited to use in thermoplastics. (See ref. 5.)

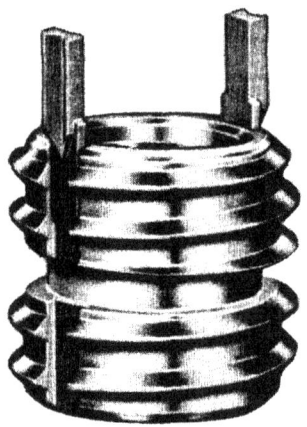

Figure 21.—Keensert.

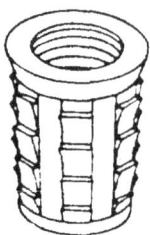

Figure 22.—Plastic expandable insert.

Figure 23.—Molded-in-place insert.

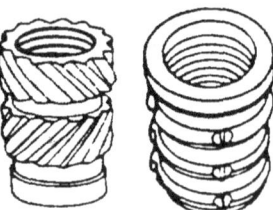

Figure 24.—Ultrasonic inserts.

Threads

Types of Threads

Since complete information on most threads can be found in the ANSI standards (ref. 6), the SAE Handbook (ref. 4), and the National Institute of Standards and Technology (formerly the National Bureau of Standards) Handbook H–28 (ref. 7) no thread standards will be included in this handbook. The goal here is to explain the common thread types, along with their advantages and disadvantages. The common thread types are unified national coarse (UNC), unified national fine (UNF), unified national extra fine (UNEF), UNJC, UNJF, UNR, UNK, and constant-pitch threads.

Unified national coarse.—UNC is the most commonly used thread on general-purpose fasteners. Coarse threads are deeper than fine threads and are easier to assemble without cross threading. The manufacturing tolerances can be larger than for finer threads, allowing for higher plating tolerances. UNC threads are normally easier to remove when corroded, owing to their sloppy fit. However, a UNC fastener can be procured with a class 3 (tighter) fit if needed (classes to be covered later).

Unified national fine.—UNF thread has a larger minor diameter than UNC thread, which gives UNF fasteners slightly higher load-carrying and better torque-locking capabilities than UNC fasteners of the same identical material and outside diameter. The fine threads have tighter manufacturing tolerances than UNC threads, and the smaller lead angle allows for finer tension adjustment. UNF threads are the most widely used threads in the aerospace industry.

Unified national extra fine.—UNEF is a still finer type of thread than UNF and is common to the aerospace field. This thread is particularly advantageous for tapped holes in hard materials and for thin threaded walls, as well as for tapped holes in thin materials.

UNJC and UNJF threads.—"J" threads are made in both external and internal forms. The external thread has a much larger root radius than the corresponding UNC, UNR, UNK, or UNF threads. This radius is mandatory and its inspection is required, whereas no root radius is required on UNC, UNF, or UNEF threads. Since the larger root radius increases the minor diameter, a UNJF or UNJC fastener has a larger net tensile area than a corresponding UNF or UNC fastener. This root radius also gives a smaller stress concentration factor in the threaded section. Therefore, high-strength ($\geq$ 180 ksi) bolts usually have "J" threads.

UNR threads.—The UNR external thread is a rolled UN thread in all respects except that the root radius must be rounded. However, the root radius and the minor diameter are *not* checked or toleranced. There is no internal UNR thread.

UNK threads.—The UNK external threads are similar to UNR threads, except that the root radius and the minor diameter *are* toleranced and inspected. There is no internal UNK thread.

According to a survey of manufacturers conducted by the Industrial Fasteners Institute, nearly all manufacturers of externally threaded fasteners make UNR rolled threads rather than plain UN. The only exception is for ground or cut threads.

Constant-pitch threads.—These threads offer a selection of pitches that can be matched with various diameters to fit a particular design. This is a common practice for bolts of 1-in. diameter and above, with the pitches of 8, 12, or 16 threads per inch being the most common.

A graphical and tabular explanation of UN, UNR, UNK, and UNJ threads is given on page M-6 of reference 8. A copy (fig. 25) is enclosed here for reference.

Classes of Threads

Thread classes are distinguished from each other by the amounts of tolerance and allowance. The designations run from 1A to 3A and 1B to 3B for external and internal threads, respectively. A class 1 is a looser fitting, general-purpose thread; a class 3 is the closer-toleranced aerospace standard thread. (The individual tolerances and sizes for the various classes are given in the SAE Handbook (ref 4).)

Forming of Threads

Threads may be cut, hot rolled, or cold rolled. The most common manufacturing method is to cold form both the head and the threads for bolts up to 1 in. in diameter. For bolts above 1-in. diameter and high-strength smaller bolts, the heads are hot forged. The threads are still cold rolled until the bolt size prohibits the material displacement necessary to form the threads (up to a constant pitch of eight threads per inch). Threads are cut only at assembly with taps and dies or by lathe cutting.

Cold rolling has the additional advantage of increasing the strength of the bolt threads through the high compressive surface stresses, similar to the effects of shot peening. This process makes the threads more resistant to fatigue cracking.

Fatigue-Resistant Bolts

If a bolt is cycled in tension, it will normally break near the end of the threaded portion because this is the area of maximum stress concentration. In order to lessen the stress concentration factor, the bolt shank can be machined down to the root diameter of the threads. Then it will survive tensile cyclic loading much longer than a standard bolt with the shank diameter equal to the thread outside diameter.

Fatigue (Cyclic) Loading of Bolts

The bolted joint in figure 26 (from ref. 9) is preloaded with an initial load F_i, which equals the clamping load F_c, before the external load F_e is applied. The equation (from ref. 11) for this assembly is

$$F_b = F_i + \left(\frac{K_b}{K_b + K_c} \right) F_e$$

where F_b is the total bolt load. In this equation K_b is the spring constant of the bolt and K_c is the spring constant of the clamped faces. To see the effects of the relative spring constants, let $R = K_c / K_b$. Then (from ref. 10)

$$F_b = F_i + \left(\frac{1}{1 + R} \right) F_e$$

In a normal clamped joint K_c is much larger than K_b ($R \approx 5.0$ for steel bolt and flanges), so that the bolt load does not increase much as the initial external load F_e is applied. (Note that the bolt load does not increase significantly until F_e exceeds F_i.)

In order to further clarify the effect of externally applied loads, a series of triangular diagrams (fig. 27, from ref. 11) can be used to illustrate loading conditions.

Triangle OAB is identical in all four diagrams. The slope of OA represents the bolt stiffness; the slope of AB represents the joint stiffness (joint is stiffer than bolt by ratio OC/CB.) In figure 27(a) the externally applied load $F_e(a)$ does not load the bolt to its yield point. In figure 27(b) the bolt is loaded by $F_e(b)$ to its yield point, with the corresponding decrease in clamping load to F_{CL}. In figure 27(c) external load $F_e(c)$ has caused the bolt to take a permanent elongation such that the clamping force will be less than F_i when $F_e(c)$ is removed. In figure 27(d) the joint has completely separated on its way to bolt failure.

Note that the flatter the slope of OA (or the larger the ratio OC/OB becomes), the smaller the effect F_e has on bolt load. Therefore, using more smaller-diameter fasteners rather than a few large-diameter fasteners will give a more fatigue-resistant joint.

Referring to figure 27(a), note that the cyclic (alternating) load is that portion above F_i. This is the alternating load

60° SCREW THREAD NOMINAL FORMS (SEE ANSI STANDARDS FOR FURTHER DETAILS)

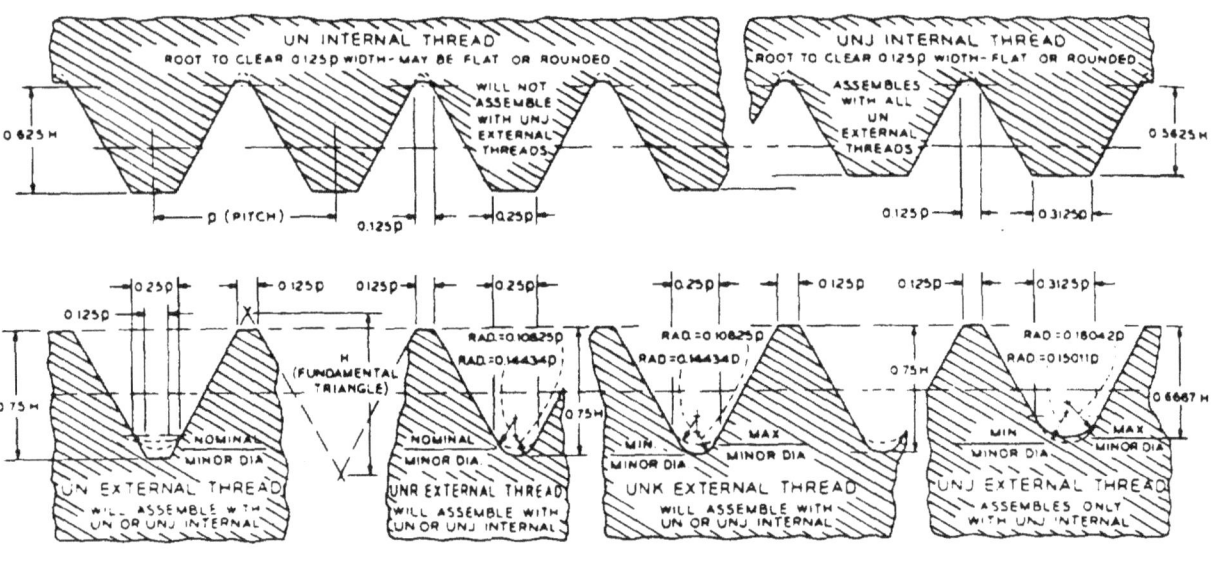

THREAD IDENTIFICATION	UN THREADS	UNR THREADS	UNK THREADS	UNJ THREADS
	Internal and External	External Only	External Only	Internal and External
ANSI[1] STANDARDS DOCUMENTS	Unified Screw Threads B1.1–1960 (See Page M–7) Metric Translation B1.1a–1968 Gages and Gaging for Unified Screw Threads B1.2–1966	Unified Screw Threads B1.1–1960 See Page M–7) Metric Translation B1.1a–1968 (Draft) UNR Addendum to B1.1–1960 (See Page M–19) Gages and Gaging for Unified Screw Threads B1.2–1966	(Draft) B1.14 for Form and Conformance	(Draft) B1.15 for Form and Conformance (No Radius Required on Internal Thread)
EXTERNAL ROOT	External Thread Root may be Flat or Rounded	External Thread Root Radius Required	External Thread Root Radius Mandatory Check Required	External Thread Root Radius Mandatory Check Required
EXTERNAL MINOR DIAMETER	External Thread Minor Diameter is not Toleranced	External Thread Minor Diameter is not Toleranced	External Thread Minor Diameter is Toleranced	External Thread Minor Diameter is Toleranced
EXTERNAL THREADS	UN Classes 1A, 2A and 3A	UNR Classes 1A, 2A and 3A	UNK Classes 2A and 3A	UNJ Class 3A Mates only with UNJ Internal Threads
INTERNAL THREADS	UN Classes 1B, 2B and 3B	No Internal Threads Designated UNR UNR Mates with UN Internal Thread	No Internal Threads Designated UNK Mates with UN or UNJ Internal Thread	UNJ Classes 3B and 3BG (No Radius Required on Internal Thread)
ANGLE AND LEAD TOLERANCE	Individually Equivalent to 50% of P.D. Tolerance Checked only when Specified	Individually Equivalent to 50% of P.D. Tolerance Checked only when Specified	Individually Equivalent to 40% of P.D. Tolerance Mandatory Check Required	Individually Equivalent to 40% of P.D. Tolerance Mandatory Check Required

NOTES: 1 Refer to the appropriate Standards, as listed, for complete thread details and conformance data. The appropriate current Standard is the authoritative document for complete details and data, and takes precedence over this sheet.

2 These Standards may be obtained through ASME

Figure 25.—Explanation of UN, UNR, UNK, and UNJ threads. (From ref. 8.) Reprinted with permission of Industrial Fasteners Institute.

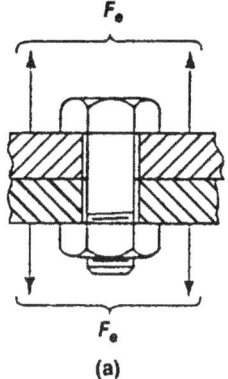 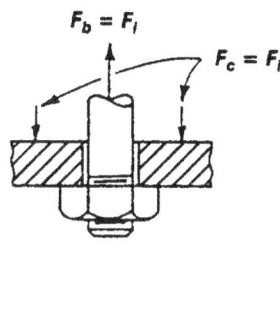 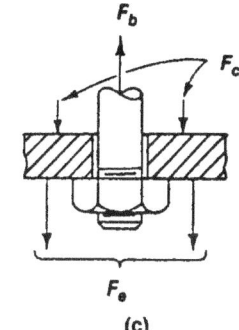

(a) Bolted flanges with external load.
(b) Free body with no external load.
(c) Free body with external load.

Figure 26.—Fatigue loading of bolts.

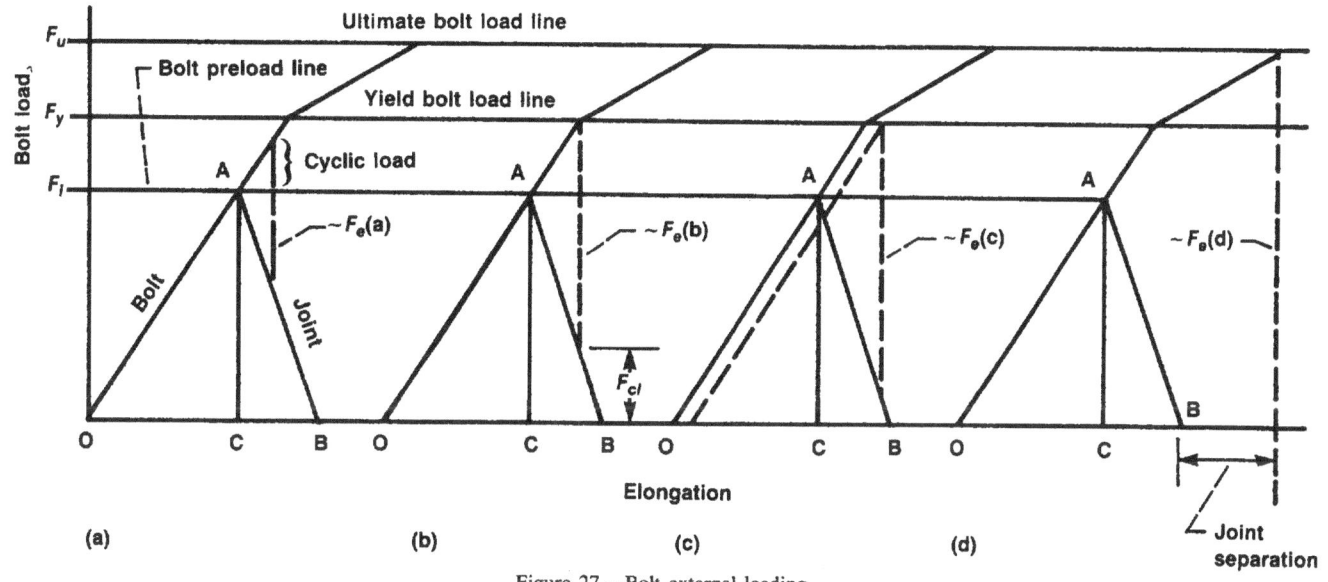

Figure 27.—Bolt external loading.

(stress) to be used on a stress-versus-load-cycles diagram of the bolt material to predict the fatigue life of the bolts. Note that an initial preload F_i near the bolt yields minimizes cyclic loading.

Thermal Cyclic Loading of Bolts

If the bolt and joint are of different materials, an operating temperature higher or lower than the installation temperature can cause problems. Differential contraction can cause the joint to unload (or separate); differential expansion can cause overloading of the fasteners. In these cases it is common practice to use conical washers (see washer section of this manual) to give additional adjustments in fastener and joint loading.

Fastener Torque

Determining the proper torque for a fastener is the biggest problem in fastener installation. Some of the many variables causing problems are
  (1) The coefficient of friction between mating threads
  (2) The coefficient of friction between the bolthead (or nut) and its mating surface
  (3) The effect of bolt coatings and lubricants on the friction coefficients
  (4) The percentage of bolt tensile strength to be used for preload
  (5) Once agreement is reached on item 4, how to accurately determine this value
  (6) Relative spring rates of the structure and the bolts

TABLE IV.—COEFFICIENTS OF STATIC AND SLIDING FRICTION
[From ref. 12.]

Materials	Static Dry	Static Greasy	Sliding Dry	Sliding Greasy
Hard steel on hard steel	0.78(1)	0.11(1,a) 0.23(1,b) 0.15(1,c) 0.11(1,d) 0.0075(18,p) 0.0052(18,h)	0.42(2)	0.029(5,k) 0.081(5,e) 0.080(5,i) 0.058(5,j) 0.084(5,d) 0.105(5,k) 0.096(5,l) 0.108(5,m) 0.12(5,a)
Mild steel on mild steel	0.74(19)		0.57(3)	0.09(3,a) 0.19(3,u)
Hard steel on graphite	0.21(1)	0.09(1,a)		
Hard steel on babbitt (ASTM No. 1)	0.70(11)	0.23(1,b) 0.15(1,c) 0.08(1,d) 0.085(1,e)	0.33(6)	0.16(1,b) 0.06(1,c) 0.11(1,d)
Hard steel on babbitt (ASTM No. 8)	0.42(11)	0.17(1,b) 0.11(1,c) 0.09(1,d) 0.08(1,e)	0.35(11)	0.14(1,b) 0.065(1,c) 0.07(1,d) 0.08(11,h)
Hard steel on babbitt (ASTM No. 10)		0.25(1,b) 0.12(1,c) 0.10(1,d) 0.11(1,e)		0.13(1,b) 0.06(1,c) 0.055(1,d)
Mild steel on cadmium silver				0.097(2,f)
Mild steel on phosphor bronze			0.34(3)	0.173(2,f)
Mild steel on copper lead		0.183(15,c)	0.23(6)	0.145(2,f)
Mild steel on cast iron				0.133(2,f)
Mild steel on lead	0.95(11)	0.5(1,f)	0.95(11)	0.3(11,f)
Nickel on mild steel	0.64(3)			0.178(3,x)
Aluminum on mild steel	0.61(8)		0.47(3)	
Magnesium on mild steel			0.42(3)	
Magnesium on magnesium	0.6(22)	0.08(22,y)		
Teflon on Teflon	0.04(22)			0.04(22,f)
Teflon on steel	0.04(22)			0.04(22,f)

Materials	Static Dry	Static Greasy	Sliding Dry	Sliding Greasy
Tungsten carbide on tungsten carbide	0.2(22)	0.12(22,a)		
Tungsten carbide on steel	0.5(22)	0.08 (22,a)		
Tungsten carbide on copper	0.35(23)			
Tungsten carbide on iron	0.8(23)			
Bonded carbide on copper	0.35(23)			
Bonded carbide on iron	0.8(23)			
Cadmium on mild steel			0.46(3)	
Copper on mild steel	0.53(8)		0.36(3)	0.18(17,a)
Nickel on nickel	1.10(16)		0.53(3)	0.12(3,w)
Brass on mild steel	0.51(8)		0.44(6)	
Brass on cast iron			0.30(6)	
Zinc on cast iron	0.85(16)		0.21(7)	
Magnesium on cast iron			0.25(7)	
Copper on cast iron	1.05(16)		0.29(7)	
Tin on cast iron			0.32(7)	
Lead on cast iron			0.43(7)	
Aluminum on aluminum	1.05(16)		1.4(3)	
Glass on glass	0.94(8)	0.01(10,p) 0.005(10,q)	0.40(3)	0.09(3,a) 0.116(3,v)
Carbon on glass			0.18(3)	
Garnet on mild steel			0.39(3)	
Glass on nickel	0.78(8)		0.56(3)	
Copper on glass	0.68(8)		0.53(3)	
Cast iron on cast iron	1.10(16)		0.15(9)	0.070(9,d) 0.064(9,n)
Bronze on cast iron			0.22(9)	0.77(9,n)
Oak on oak (parallel to grain)	0.62(9)		0.48(9)	0.164(9,r) 0.067(9,s)
Oak on oak (perpendicular)			0.32(9)	0.072(9,s)
Leather on oak (parallel)	0.54(9)		0.52(9)	0.075(9,n)
Cast iron on oak	0.61(9)		0.49(9)	0.36(9,t)
Leather on cast iron			0.56(9)	0.13(9,n)
Laminated plastic on steel			0.35(12)	0.05(12,f)
Fluted rubber bearing on steel				0.05(13,t)

(1) Campbell, *Trans. ASME*, 1939; (2) Clarke, Lincoln, and Sterrett, *Proc. API*, 1935; (3) Beare and Bowden, *Phil. Trans. Roy. Soc.*, 1935; (4) Dokos, *Trans. ASME*, 1946; (5) Boyd and Robertson, *Trans. ASME*, 1945; (6) Sachs, *zeit. f. angew. Math. and Mech.*, 1924; (7) Honda and Yamada, *Jour. I of M.* 1925; (8) Tomlinson, *Phil. Mag.*, 1929; (9) Morin, *Acad. Roy. des Sciences*, 1838; (10) Claypoole, *Trans. ASME*, 1943; (11) Tabor, *Jour. Applied Phys.*, 1945; (12) Eyssen, General Discussion on Lubrication, *ASME*, 1937; (13) Brazier and Holland-Bowyer, General Discussion on Lubrication, *ASME*, 1937; (14) Burwell, *Jour. SAE*, 1942; (15) Stanton, "Friction", Longmans; (16) Ernst and Merchant, Conference on Friction and Surface Finish, M.I.T., 1940; (17) Gongwer, Conference on Friction and Surface Finish, M.I.T., 1940; (18) Hardy and Bircumshaw, *Proc. Roy. Soc.*, 1925; (19) Hardy and Hardy, *Phil. Mag.*, 1919; (20) Bowden and Young, *Proc. Roy. Soc.*, 1951; (21) Hardy and Doubleday, *Proc. Roy. Soc.*, 1923; (22) Bowden and Tabor, "The Friction and Lubrication of Solids." Oxford; (23) Shooter, *Research*, 4, 1951.

(a) Oleic acid; (b) Atlantic spindle oil (light mineral); (c) castor oil; (d) lard oil; (e) Atlantic spindle oil plus 2 percent oleic acid; (f) medium mineral oil; (g) medium mineral oil plus ½ percent oleic acid; (h) stearic acid; (i) grease (zinc oxide base); (j) graphite; (k) turbine oil plus 1 percent graphite; (l) turbine oil plus 1 percent stearic acid; (m) turbine oil (medium mineral); (n) olive oil; (p) palmitic acid; (q) ricinoleic acid; (r) dry soap; (s) lard; (t) water; (u) rape oil; (v) 3-in-1 oil; (w) octyl alcohol; (x) triolein; (y) 1 percent lauric acid in paraffin oil.

(7) Interaction formulas to be used for combining simultaneous shear and tension loads on a bolt (Should friction loads due to bolt clamping action be included in the interaction calculations?)

(8) Whether "running torque" for a locking device should be added to the normal torque

Development of Torque Tables

The coefficient of friction can vary from 0.04 to 1.10, depending on the materials and the lubricants being used between mating materials. (Table IV from ref. 12 gives a variety of friction coefficients.) Since calculated torque values are a function of the friction coefficients between mating threads and between the bolthead or nut and its mating surface, it is vitally important that the torque table values used are adjusted to reflect any differences in friction coefficients between those used to calculate the table and the user's values. Running torque should be included in the values listed in the tables because any torque puts shear load on the bolt.

The torque values in table V have been calculated as noted in the footnotes, by using formulas from reference 13. (A similar table was published in *Product Engineering* by Arthur Korn around 1944.)

Higher torques (up to theoretical yield) are sometimes used for bolts that cannot be locked to resist vibration. The higher load will increase the vibration resistance of the bolt, but the bolt will yield and unload if its yield point is inadvertently exceeded. Since the exact yield torque cannot be determined without extensive instrumentation, it is not advisable to torque close to the bolt yield point.

Fastener proof load is sometimes listed in the literature. This value is usually 75 percent of theoretical yield, to prevent inadvertent yielding of the fastener through torque measurement inaccuracies.

Alternative Torque Formula

A popular formula for quick bolt torque calculations is $T = KFd$, where T denotes torque, F denotes axial load, d denotes bolt diameter, and K (torque coefficient) is a calculated value from the formula:

$$K = \left(\frac{d_m}{2d}\right) \frac{\tan \psi + \mu \sec \alpha}{1 - \mu \tan \psi \sec \alpha} + 0.625 \mu_c$$

as given in reference 14 (p. 378) where

d_m thread mean diameter

ψ thread helix angle

μ friction coefficient between threads

α thread angle

μ_c friction coefficient between bolthead (or nut) and clamping surface

The commonly assumed value for K is 0.2, but this value should not be used blindly. Table VI gives some calculated values of K for various friction coefficients. A more realistic "typical" value for K would be 0.15 for steel on steel. Note that μ and μ_c are not necessarily equal, although equal values were used for the calculated values in table VI.

Torque-Measuring Methods

A number of torque-measuring methods exist, starting with the mechanic's "feel" and ending with installing strain gages on the bolt. The accuracy in determining the applied torque values is cost dependent. Tables VII and VIII are by two different "experts," and their numbers vary. However, they both show the same trends of cost versus torque accuracy.

Design Criteria

Finding Shear Loads on Fastener Group

When the load on a fastener group is eccentric, the first task is to find the centroid of the group. In many cases the pattern will be symmetrical, as shown in figure 28. The next step is to divide the load R by the number of fasteners n to get the direct shear load P_c (fig. 29(a)). Next, find Σr_n^2 for the group of fasteners, where r_n is the radial distance of each fastener from the centroid of the group. Now calculate the moment about the centroid ($M = Re$ from fig. 28). The contributing shear load for a particular fastener due to the moment can be found by the formula

$$P_e = \frac{Mr}{\Sigma r_n^2}$$

where r is the distance (in inches) from the centroid to the fastener in question (usually the outermost one). Note that this is analogous to the torsion formula, $f = Tr/J$, except that P_e is in pounds instead of stress. The two loads (P_c and P_e) can now be added vectorally as shown in figure 29(c) to get the resultant shear load P (in pounds) on each fastener. Note that the fastener areas are all the same here. If they are unequal, the areas must be weighted for determining the centroid of the pattern.

Further information on this subject may be found in references 16 and 17.

Finding Tension Loads on Fastener Group

This procedure is similar to the shear load determination, except that the centroid of the fastener group may not be the geometric centroid. This method is illustrated by the bolted bracket shown in figure 30.

The pattern of eight fasteners is symmetrical, so that the tension load per fastener from P_1 will be $P_1/8$. The additional

TABLE V.—BOLT TORQUE

[No lubrication on threads. Torque values are based on friction coefficients of 0.12 between threads and 0.14 between nut and washer or head and washer, as manufactured (no special cleaning).]

Size	Root area, in.^{2}	Torque range (class 8, 150 ksi, bolts[a])
10–24	0.0145	23 to 34 in.-lb
10–32	.0175	29 to 43 in.-lb
¼–20	.0269	54 to 81 in.-lb
¼–4–28	.0326	68 to 102 in.-lb
$^5/_{16}$–18	.0454	117 to 176 in.-lb
$^5/_{16}$–24	.0524	139 to 208 in.-lb
⅜–16	.0678	205 to 308 in.-lb
⅜–24	.0809	230 to 345 in.-lb
$^7/_{16}$–14	.0903	28 to 42 ft-lb
$^7/_{16}$–20	.1090	33 to 50 ft-lb
½–13	.1257	42 to 64 ft-lb
½–20	.1486	52 to 77 ft-lb
$^9/_{16}$–12	.1620	61 to 91 ft-lb
$^9/_{16}$–18	.1888	73 to 109 ft-lb
⅝–11	.2018	84 to 126 ft-lb
⅝–18	.2400	104 to 156 ft-lb
¾–10	.3020	[b]117 to 176 ft-lb
¾–16	.3513	[b]139 to 208 ft-lb
⅞–9	.4193	[b]184 to 276 ft-lb
⅞–14	.4805	[b]213 to 320 ft-lb
1–8	.5510	[b]276 to 414 ft-lb
1–14	.6464	[b]323 to 485 ft-lb
1⅛–7	.6931	[b]390 to 585 ft-lb
1⅛–12	.8118	[b]465 to 698 ft-lb
1¼–7	.8898	[b]559 to 838 ft-lb
1¼–12	1.0238	[b]655 to 982 ft-lb

[a]The values given are 50 and 75 percent of theoretical yield strength of a bolt material with a yield of 120 ksi. Corresponding values for materials with different yield strengths can be obtained by multiplying these table values by the ratio of the respective material yield strengths.

[b]Bolts of 0.75-in. diameter and larger have reduced allowables (75 percent of normal strength) owing to inability to heat treat this large a cross section to an even hardness.

Reprinted from Machine Design, Nov. 19, 1987. Copyright, 1987 by Penton Publishing, Inc., Cleveland, OH.

TABLE VI.—TORQUE COEFFICIENTS

Friction coefficient		Torque coefficient, K
Between threads, μ	Between bolthead (or nut) and clamping surface, μ_c	
0.05	0.05	0.074
.10	.10	.133
.15	.15	.189
.20	.20	.250

TABLE VII.—INDUSTRIAL FASTENERS INSTITUTE'S TORQUE-MEASURING METHOD

[From ref. 8.]

Preload measuring method	Accuracy, percent	Relative cost
Feel (operator's judgment)	±35	1
Torque wrench	±25	1.5
Turn of the nut	±15	3
Load-indicating washers	±10	7
Fastener elongation	±3 to 5	15
Strain gages	±1	20

moment P_2h will also produce a tensile load on some fasteners, but the problem is to determine the "neutral axis" line where the bracket will go from tension to compression. If the plate is thick enough to take the entire moment P_2h in bending at the edge AB, that line could be used as the heeling point, or neutral axis. However, in this case, I have taken the conservative approach that the plate will not take the bending and will heel at the line CD. Now the Σr_n^2 will only include bolts 3 to 8, and the r_n's (in inches) will be measured from line CD. Bolts 7 and 8 will have the highest tensile loads (in pounds), which will be $P = P_T + P_M$, where $P_T = P_1/8$ and

$$P_m = \frac{Mr}{\Sigma r_n^2} = \frac{P_2 h r_7}{\Sigma r_n^2}$$

An alternative way of stating this relationship is that the bolt load is proportional to its distance from the pivot axis and the moment reacted is proportional to the sum of the squares of the respective fastener distances from the pivot axis.

At this point the applied total tensile load should be compared with the total tensile load due to fastener torque. The torque should be high enough to exceed the maximum applied tensile load in order to avoid joint loosening or leaking. If the bracket geometry is such that its bending capability cannot be readily determined, a finite element analysis of the bracket itself may be required.

Combining Shear and Tensile Fastener Loads

When a fastener is subjected to both tensile and shear loading simultaneously, the combined load must be compared with the total strength of the fastener. Load ratios and interaction curves are used to make this comparison. The load ratios are

$$R_S(\text{or } R_1) = \frac{\text{Actual shear load}}{\text{Allowable shear load}}$$

$$R_T(\text{or } R_2) = \frac{\text{Actual tensile load}}{\text{Allowable tensile load}}$$

TABLE VIII.—MACHINE DESIGN'S TORQUE-MEASURING METHOD

[From ref. 15.]

(a) Typical tool accuracies

Type of tool	Element controlled	Typical accuracy range, percent of full scale
Slug wrench	Turn	1 Flat
Bar torque wrench	Torque	±3 to 15
	Turn	1/4 Flat
Impact wrench	Torque	±10 to 30
	Turn	±10 to 20°
Hydraulic wrench	Torque	±3 to ±10
	Turn	±5 to 10°
Gearhead air-powered wrench	Torque	±10 to ±20
	Turn	±5 to 10°
Mechanical multiplier	Torque	±5 to 20
	Turn	±2 to 10°
Worm-gear torque wrench	Torque	±0.25 to 5
	Turn	±1 to 5°
Digital torque wrench	Torque	±1/4 to 1
	Turn	1/4 Flat
Ultrasonically controlled wrench	Bolt elongation	±1 to 10
Hydraulic tensioner	Initial bolt stretch	±1 to 5
Computer-controlled tensioning	Simultaneous torque and turn	±0.5 to 2

(b) Control accuracies

Element controlled	Preload accuracy, percent	To maximize accuracy
Torque	±15 to ±30	Control bolt, nut, and washer hardness, dimensions, and finish. Have consistent lubricant conditions, quantities, application, and types.
Turn	±15 to ±30	Use consistent snug torque. Control part geometry and finish. Use new sockets and fresh lubes.
Torque and turn	±10 to ±25	Plot torque vs turn and compare to previously derived set of curves. Control bolt hardness, finish, and geometry.
Torque past yield	±3 to ±10	Use "soft" bolts and tighten well past yield point. Use consistent snugging torque. Control bolt hardness and dimensons.
Bolt stretch	±1 to ±8	Use bolts with flat, parallel ends. Leave transducer engaged during tightening operation. Mount transducer on bolt centerline.

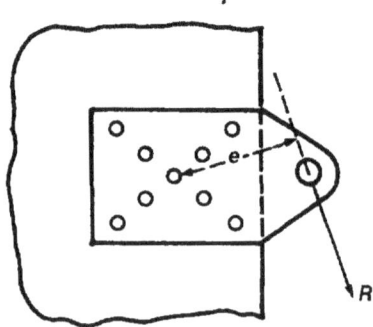

Figure 28.—Symmetrical load pattern.

The interaction curves of figure 31 are a series of curves with their corresponding empirical equations. The most conservative is $R_1 + R_2 = 1$ and the least conservative is $R_1^3 + R_2^3 = 1$. This series of curves is from an old edition of MIL–HDBK–5. It has been replaced by a single formula, $R_S^3 + R_T^2 = 1$, in the latest edition (ref. 18). However, it is better to use $R_T + R_S = 1$ if the design can be conservative with respect to weight and stress.

Note that the interaction curves do not take into consideration the friction loads from the clamped surfaces in arriving at bolt shear loads. In some cases the friction load could reduce the bolt shear load substantially.

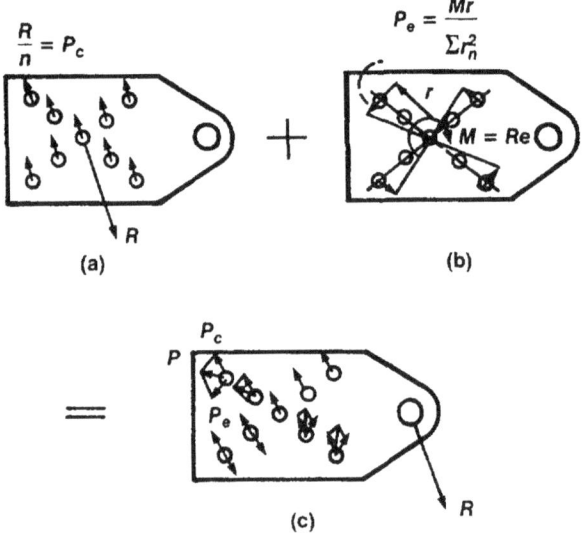

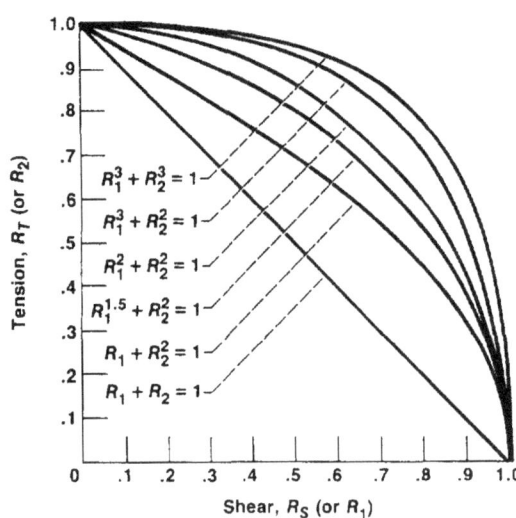

Figure 29.—Combining of shear and moment loading.

Figure 31.—Interaction curves.

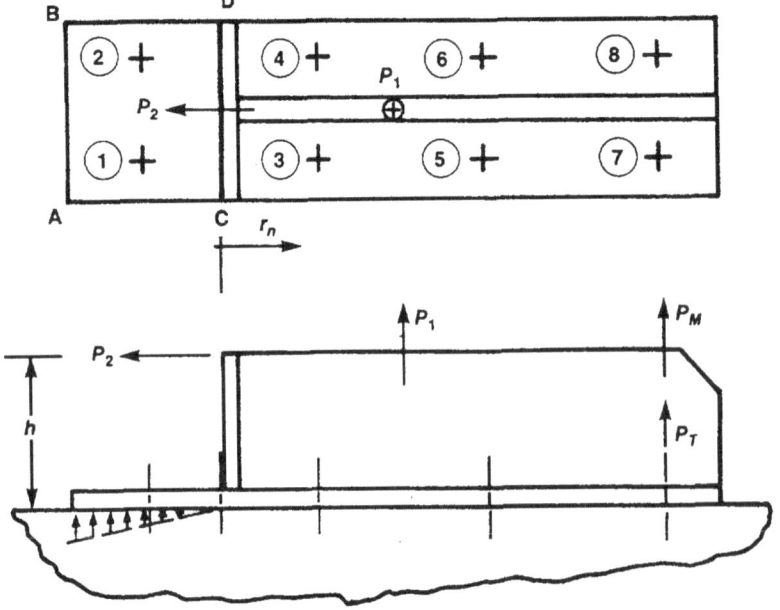

Figure 30.—Bolted bracket.

20

The margin of safety[12] for a fastener from figure 31 is

$$MS = \frac{1}{R_S^x + R_T^y} - 1$$

depending on which curve is used. However, note that $R_S^x + R_T^y < 1$ is a requirement for a positive margin of safety. This formula also illustrates why high torque should not be applied to a bolt when the dominant load is shear.

The margin of safety is calculated for *both* yield and ultimate material allowables, with the most critical value controlling the design. A material with a low yield will be critical for yield stress, and a material with a high yield will normally be critical for ultimate stress.

Calculating Pullout Load for Threaded Hole

In many cases a bolt of one material may be installed in a tapped hole in a different (and frequently lower strength) material. If the full strength of the bolt is required, the depth of the tapped hole must be determined for the weaker material by using the formula

$$P = \frac{\pi d_m F_s L}{3}$$

where

P pullout load, lb

d_m mean diameter of threaded hole, in. ($\approx$ pitch diameter of threads)

F_s material ultimate or yield shear stress

L length of thread engagement, in.

The ⅓ factor is empirical. If the threads were perfectly mated, this factor would be ½, since the total cylindrical shell area of the hole would be split equally between the bolt threads and the tapped hole threads. The ⅓ is used to allow for mismatch between threads.

Further information on required tapped hole lengths is given in reference 19.

Calculating Shank Diameter for "Number" Fastener

The shank diameter for a "number" fastener is calculated from

$$\text{Diameter} = 0.060 + 0.013\ N$$

---

where N is the number (4, 6, 8, 10, 12) of the fastener. For example, the shank diameter of a no. 8 fastener is

$$\text{Diameter} = 0.060 + 0.013(8) = 0.164 \text{ in.}$$

Fastener Groups in Bearing (Shear Loading)

Whenever possible, bolts in shear should have a higher shear strength than the bearing yield strength of the materials they go through. Since the bolts have some clearance and position tolerances in their respective holes, the sheet material must yield in bearing to allow the bolt pattern to load all of the bolts equally at a given location in the pattern. Note that the sloppier the hole locations, the more an individual bolt must carry before the load is distributed over the pattern.

Bolts and rivets should not be used together to carry a load, since the rivets are usually installed with an interference fit. Thus, the rivets will carry all of the load until the sheet or the rivets yield enough for the bolts to pick up some load. This policy also applies to bolts and dowel pins (or roll pins) in a pattern, since these pins also have interference fits.

Fastener Edge Distance and Spacing

Common design practice is to use a nominal edge distance of $2D$ from the fastener hole centerline, where D is the fastener diameter. The minimum edge distance should not be less than $1.5D$. The nominal distance between fasteners is $4D$, but the thickness of the materials being joined can be a significant factor. For thin materials, buckling between fasteners can be a problem. A wider spacing can be used on thicker sheets, as long as sealing of surfaces between fasteners is not a problem.

Approximate Bearing and Shear Allowables

In the absence of specific shear and bearing allowables for materials, the following approximations may be used:

Alloy and carbon steels: $F_{su} = 0.6\ F_{tu}$

Stainless steels: $F_{su} = 0.55\ F_{tu}$

where F_{su} is ultimate shear stress and F_{tu} is ultimate tensile stress. Since bearing stress allowables are empirical to begin with, the bearing allowable for any given metallic alloy may be approximated as follows:

$$F_{bu} = 1.5\ F_{tu}$$

$$F_{by} = 1.5\ F_{ty}$$

where F_{bu} is ultimate bearing stress, F_{by} is yield bearing stress, and F_{ty} is tensile yield stress.

Proper Fastener Geometry

Most military standard (MS) and national aerospace standard (NAS) fasteners have coded callouts that tell the diameter, grip length, drilling of the head or shank, and the material (where the fastener is available in more than one material). Rather than listing a group of definitions, it is easier to use the NAS 1003 to NAS 1020 (fig. 32) as an example to point out the following:

(1) The last two digits give the fastener diameter in sixteenths of an inch.

(2) The first dash number is the grip length in sixteenths of an inch.

(3) The letters given with the dash number indicate the head and/or shank drilling.

In addition, an identifying letter or dash number is added to indicate the fastener material. However, this systematic practice is not rigidly followed in all MS and NAS fastener standards.

Shear Heads and Nuts

In the aerospace industry the general ground rule is to design such that fasteners are primarily in shear rather than tension. As a result, many boltheads and nuts are made about one-half as thick as normal to save weight. These bolts and nuts are referred to as shear bolts and shear nuts, and care must be used in never specifying them for tension applications. The torque table values must also be reduced to one-half for these bolts and nuts.

Use of Proper Grip Length

Standard design practice is to choose a grip length such that the threads are never in bearing (shear). Where an exact grip length is not available, the thickness of the washers used under the nut or bolthead can be varied enough to allow proper grip.

Bolthead and Screwhead Styles

Although the difference between bolts and screws is not clearly defined by industry, at least the head styles are fairly well defined. The only discrepancy found in figure 33 is that the plain head, with a square shoulder, is more commonly called a carriage bolthead. The angle of countersunk heads (flat) can vary from 60° to 120°, but the common values are 82° and 100°.

Counterfeit Fasteners

In the past two years a great deal of concern and publicity about counterfeit fasteners has surfaced. The counterfeit case with the most documentation is the deliberate marking of grade 8.2 boron bolts as grade 8 bolts.

Grade 8.2 bolts are a low-carbon (0.22 percent C) boron alloy steel that can be heat treated to the same room-temperature hardness as grade 8 medium-carbon (0.37 percent C) steel. However, the room- and elevated-temperature strengths of the grade 8.2 bolts drop drastically if they are exposed to temperatures above 500 °F. Grade 8 bolts can be used to 800 °F with little loss of room-temperature strength.

Other fasteners marked as MS and NAS but not up to the respective MS or NAS specification have shown up; however, documentation is not readily available. Since these fasteners are imported and have no manufacturer's identification mark on them, it is not possible to trace them back to the guilty manufacturer. U.S. Customs inspections have not been effective in intercepting counterfeit fasteners.

Another problem with fasteners has been the substitution of zinc coating for cadmium coating. If a dye is used with the zinc, the only way to detect the difference in coatings is by chemical testing.

Federal legislation to establish control of fastener materials from the material producer to the consumer is being formulated.

Bolthead Identification

Identifying an existing non-MS, non-NAS, or non-Air Force-Navy bolt is usually a problem. Each manufacturer seems to have a different system. Frank Akstens of Fastener Technology International magazine (ref. 20) has compiled a good listing of several hundred "common" bolts. His entire compilation is enclosed as appendix A of this report. An international guide to bolt manufacturer's identification symbols has also been published by Fastener Technology International magazine.

Fastener Strength

Allowable strengths for many types of fasteners are given in MIL-HDBK-5 (ref. 18). Ultimate shear and tensile strengths of various threaded fasteners are given in appendix B of this report.

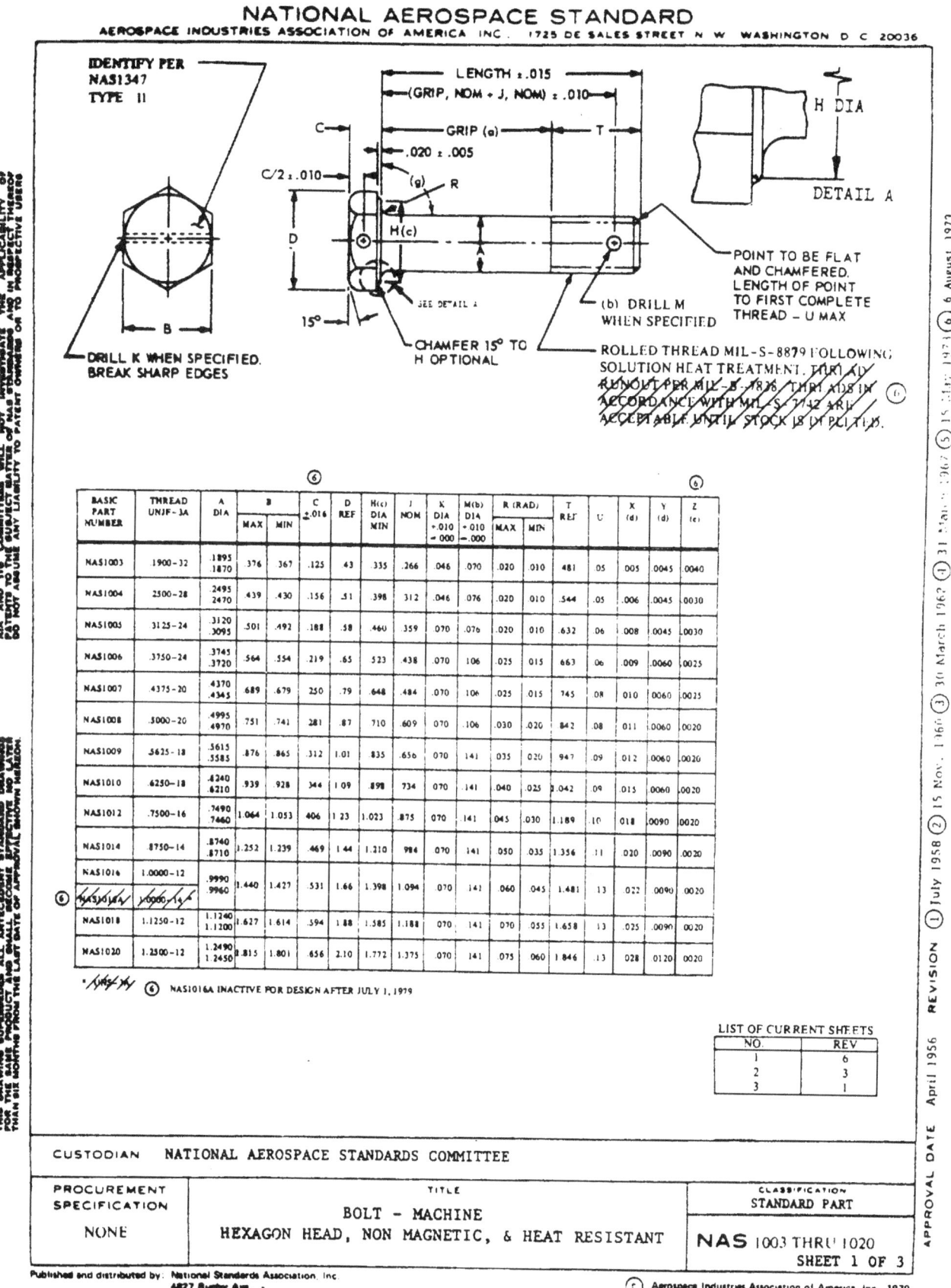

Figure 32.—National aerospace standard for proper fastener geometry.

CODE: BASIC PART NUMBER DESIGNATES NOMINAL DIAMETER.
         DASH NUMBER DESIGNATES GRIP AND LENGTH (SEE SHEET 3).
         ADD "A" TO DASH NUMBER FOR UNDRILLED BOLT.
         ADD "H" TO DASH NUMBER FOR DRILLED HEAD ONLY.
         NO CODE LETTER DESIGNATES DRILLED SHANK ONLY.

EXAMPLE: NAS1003-8 = .1900 DIAMETER BOLT, .500 GRIP, DRILLED SHANK ONLY.
             NAS1003-8A = .1900 DIAMETER BOLT, .500 GRIP, UNDRILLED.
             NAS1003-8H = .1900 DIAMETER BOLT, .500 GRIP, DRILLED HEAD ONLY.

MATERIAL: CRES, A-286 SPEC AMS5735 OR AMS5737 OR DISCALOY SPEC AMS5753, EXCEPT ULTIMATE ③
             TENSILE STRENGTH 140,000 PSI MINIMUM AT ROOM TEMPERATURE, FABRICATED TO AMS7478.

FINISH: CLEAN AND PASSIVATE IN ACCORDANCE WITH MIL-S-5002 QQ-P-35   ①

NOTES: 1. REFERENCE DIMENSIONS ARE FOR DESIGN PURPOSES ONLY AND NOT AN INSPECTION REQUIREMENT.
        2. MAGNETIC PERMEABILITY SHALL BE LESS THAN 2.0 (AIR = 1.0) FOR A FIELD STRENGTH H = 200 OERSTEDS
            (MAGNETIC PERMEABILITY INDICATOR PER MIL-I-17214 OR EQUIVALENT.)
        3. BOLTS SHALL BE FREE FROM BURRS AND SLIVERS.
        4. THESE BOLTS ARE INTENDED FOR USE AT TEMPERATURES UP TO 1200 F.
    (a) 5. GRIP LENGTH: FROM UNDER SIDE OF HEAD TO END OF FULL CYLINDRICAL PORTION OF SHANK.
    (b) 6. COTTER PIN HOLE CENTERLINE: WITHIN .010 AND NORMAL WITHIN 2° OF BOLT CENTERLINE.
    (c) 7. "H" DIA MAXIMUM NOT TO EXCEED "B", MINIMUM DIA AT TOP OF HEAD NOT LESS THAN "H".
    (d) 8. CONCENTRICITY: "H" AND "A" DIAMETER WITHIN "X" VALUES TIR, "A" AND THREAD PITCH DIAMETER
            WITHIN "Y" VALUES TIR.
    (e) 9. SHANK STRAIGHTNESS: WITHIN "Z" VALUES TIR PER INCH OF LENGTH.
    (g)10. BEARING SURFACE SQUARENESS: WITHIN .005 TIR WITH SHANK.
        11. DIMENSIONS IN INCHES.
            TOLERANCES UNLESS OTHERWISE SPECIFIED: ANGLES ±5°

    ③   DISCALOY INACTIVE FOR DESIGN AFTER JULY 1, 1979

Published and distributed by   National Standards Association, Inc.
                               4827 Rugby Ave
                               Washington, D. C. 20014

APPROVAL DATE April 1956 REVISION ① 10 March 1962 ② 15 May 1968 ① 6 August 1979

NAS	1003 THRU 1020
	SHEET 2

Figure 32.—Continued.

NATIONAL AEROSPACE STANDARD

AEROSPACE INDUSTRIES ASSOCIATION OF AMERICA, INC., 1725 DE SALES STREET N.W. WASHINGTON, D.C. 20036

DASH NO.	GRIP ±.015	NAS1003	NAS1004	NAS1005	NAS1006	NAS1007	NAS1008	NAS1009	NAS1010	NAS1012	NAS1014	NAS1016 / NAS1016A	NAS1018	NAS1020
1	.062	.543	.606	.644	.725	.807	.904	1.009	1.104	1.251	1.418	1.543	1.720	1.908
2	.125	.606	.669	.757	.788	.870	.967	1.072	1.167	1.314	1.481	1.606	1.783	1.971
3	.188	.669	.732	.820	.851	.933	1.030	1.135	1.230	1.377	1.544	1.669	1.846	2.034
4	.250	.731	.794	.882	.913	.995	1.092	1.197	1.292	1.439	1.606	1.731	1.908	2.096
5	.312	.793	.856	.944	.975	1.057	1.154	1.259	1.354	1.501	1.668	1.793	1.970	2.158
6	.375	.856	.919	1.007	1.038	1.120	1.217	1.322	1.417	1.564	1.731	1.856	2.033	2.221
7	.438	.919	.982	1.070	1.101	1.183	1.280	1.385	1.480	1.627	1.794	1.919	2.096	2.284
8	.500	.981	1.044	1.132	1.163	1.245	1.342	1.447	1.542	1.689	1.856	1.981	2.158	2.346
9	.562	1.043	1.106	1.194	1.225	1.307	1.404	1.509	1.604	1.751	1.918	2.043	2.220	2.408
10	.625	1.106	1.169	1.257	1.288	1.370	1.467	1.572	1.667	1.814	1.981	2.106	2.283	2.471
11	.688	1.169	1.232	1.320	1.351	1.433	1.530	1.635	1.730	1.877	2.044	2.169	2.346	2.534
12	.750	1.231	1.294	1.382	1.413	1.495	1.592	1.697	1.792	1.939	2.106	2.231	2.408	2.596
13	.812	1.293	1.356	1.444	1.475	1.557	1.654	1.759	1.854	2.001	2.168	2.293	2.470	2.658
14	.875	1.356	1.419	1.507	1.538	1.620	1.717	1.822	1.917	2.064	2.231	2.356	2.533	2.721
15	.938	1.419	1.482	1.570	1.601	1.683	1.780	1.885	1.980	2.127	2.294	2.419	2.596	2.784
16	1.000	1.481	1.544	1.632	1.663	1.745	1.842	1.947	2.042	2.189	2.356	2.481	2.658	2.846
17	1.062	1.543	1.606	1.694	1.725	1.807	1.904	2.009	2.104	2.251	2.418	2.543	2.720	2.908
18	1.125	1.606	1.669	1.757	1.788	1.870	1.967	2.072	2.167	2.314	2.481	2.606	2.783	2.971
19	1.188	1.669	1.732	1.820	1.851	1.933	2.030	2.135	2.230	2.377	2.544	2.669	2.846	3.034
20	1.250	1.731	1.794	1.882	1.913	1.995	2.092	2.197	2.292	2.439	2.606	2.731	2.908	3.096
21	1.312	1.793	1.856	1.944	1.975	2.057	2.154	2.259	2.354	2.501	2.668	2.793	2.970	3.158
22	1.375	1.856	1.919	2.007	2.038	2.120	2.217	2.322	2.417	2.564	2.731	2.856	3.033	3.221
23	1.438	1.919	1.982	2.070	2.101	2.183	2.280	2.385	2.480	2.627	2.794	2.919	3.096	3.284
24	1.500	1.981	2.044	2.132	2.163	2.245	2.342	2.447	2.542	2.689	2.856	2.981	3.158	3.346
25	1.562	2.043	2.106	2.194	2.225	2.307	2.404	2.509	2.604	2.751	2.918	3.043	3.220	3.408
26	1.625	2.106	2.169	2.257	2.288	2.370	2.467	2.572	2.667	2.814	2.981	3.106	3.283	3.471
27	1.688	2.169	2.232	2.320	2.351	2.433	2.530	2.635	2.730	2.877	3.044	3.169	3.346	3.534
28	1.750	2.231	2.294	2.382	2.413	2.495	2.592	2.697	2.792	2.939	3.106	3.231	3.408	3.596
29	1.812	2.293	2.356	2.444	2.475	2.557	2.654	2.759	2.854	3.001	3.168	3.293	3.470	3.658
30	1.875	2.356	2.419	2.507	2.538	2.620	2.717	2.822	2.917	3.064	3.231	3.356	3.533	3.721
31	1.938	2.419	2.482	2.570	2.601	2.683	2.780	2.885	2.980	3.127	3.294	3.419	3.596	3.784
32	2.000	2.481	2.544	2.632	2.663	2.745	2.842	2.947	3.042	3.189	3.356	3.481	3.658	3.846
34	2.125	2.606	2.669	2.757	2.788	2.870	2.967	3.072	3.167	3.314	3.481	3.606	3.783	3.971
36	2.250	2.731	2.794	2.882	2.913	2.995	3.092	3.197	3.292	3.439	3.606	3.731	3.908	4.096
38	2.375	2.856	2.919	3.007	3.038	3.120	3.217	3.322	3.417	3.564	3.731	3.856	4.033	4.221
40	2.500	2.981	3.044	3.132	3.163	3.245	3.342	3.447	3.542	3.689	3.856	3.981	4.155	4.346
42	2.625	3.106	3.169	3.257	3.288	3.370	3.467	3.572	3.667	3.814	3.981	4.106	4.283	4.471
44	2.750	3.231	3.294	3.382	3.413	3.495	3.592	3.697	3.792	3.939	4.106	4.231	4.408	4.596
46	2.875	3.356	3.419	3.507	3.538	3.620	3.717	3.822	3.917	4.064	4.231	4.356	4.533	4.846
48	3.000	3.481	3.544	3.632	3.663	3.745	3.842	3.947	4.042	4.189	4.356	4.481	4.658	4.846
50	3.125	3.606	3.669	3.757	3.788	3.870	3.967	4.072	4.167	4.314	4.481	4.606	4.783	4.971
52	3.250	3.731	3.794	3.882	3.913	3.995	4.092	4.197	4.292	4.439	4.606	4.731	4.908	5.096
54	3.375	3.856	3.919	4.007	4.038	4.120	4.217	4.322	4.417	4.564	4.731	4.856	5.033	5.221
56	3.500	3.981	4.044	4.132	4.163	4.245	4.342	4.447	4.542	4.689	4.856	4.981	5.156	5.346
58	3.625	4.106	4.169	4.257	4.288	4.370	4.467	4.572	4.667	4.814	4.981	5.106	5.283	5.471
60	3.750	4.231	4.294	4.382	4.413	4.495	4.592	4.697	4.792	4.939	5.106	5.231	5.408	5.596
62	3.875	4.356	4.419	4.507	4.538	4.620	4.717	4.822	4.917	5.064	5.231	5.356	5.533	5.721
64	4.000	4.481	4.544	4.632	4.663	4.745	4.842	4.947	5.042	5.189	5.356	5.481	5.658	5.846
66	4.125	4.606	4.669	4.757	4.788	4.870	4.967	5.072	5.167	5.314	5.481	5.606	5.783	5.971
68	4.250	4.731	4.794	4.882	4.913	4.995	5.092	5.197	5.292	5.439	5.606	5.731	5.908	6.096
70	4.375	4.856	4.919	5.007	5.038	5.120	5.217	5.322	5.417	5.564	5.731	5.856	6.033	6.221
72	4.500	4.981	5.044	5.132	5.163	5.245	5.342	5.447	5.542	5.689	5.856	5.981	6.155	6.346
74	4.625	5.106	5.169	5.257	5.288	5.370	5.467	5.572	5.667	5.814	5.981	6.106	6.283	6.471
76	4.750	5.231	5.294	5.382	5.413	5.495	5.592	5.697	5.792	5.939	6.106	6.231	6.408	6.596
78	4.875	5.356	5.419	5.507	5.538	5.620	5.717	5.822	5.917	6.064	6.231	6.356	6.533	6.721
80	5.000	5.481	5.544	5.632	5.663	5.745	5.842	5.947	6.042	6.189	6.356	6.481	6.658	6.846
82	5.125	5.606	5.669	5.757	5.788	5.870	5.967	6.072	6.167	6.314	6.481	6.606	6.783	6.971
84	5.250	5.731	5.794	5.882	5.913	5.995	6.092	6.197	6.292	6.439	6.606	6.731	6.908	7.096
86	5.375	5.856	5.919	6.007	6.038	6.120	6.217	6.322	6.417	6.564	6.731	6.856	7.033	7.221
88	5.500	5.981	6.044	6.132	6.163	6.245	6.342	6.447	6.542	6.689	6.856	6.981	7.158	7.346
90	5.625	6.106	6.169	6.257	6.288	6.370	6.467	6.572	6.667	6.814	6.981	7.106	7.283	7.471
92	5.750	6.231	6.294	6.382	6.413	6.495	6.592	6.697	6.792	6.939	7.106	7.231	7.408	7.596
94	5.875	6.356	6.419	6.507	6.538	6.620	6.717	6.822	6.917	7.064	7.231	7.356	7.533	7.721
96	6.000	6.481	6.544	6.632	6.663	6.745	6.842	6.947	7.042	7.189	7.356	7.481	7.658	7.846

DASH NO. INDICATES GRIP LENGTH IN .0625 INCREMENTS. INTERMEDIATE OR LONGER LENGTHS MAY BE ORDERED BY USE OF PROPER DASH NO.

①

APPROVAL DATE May 1973   REVISION ① 6 August 1979

NAS 1003 THRU 1020
SHEET 3

Figure 32.—Concluded.

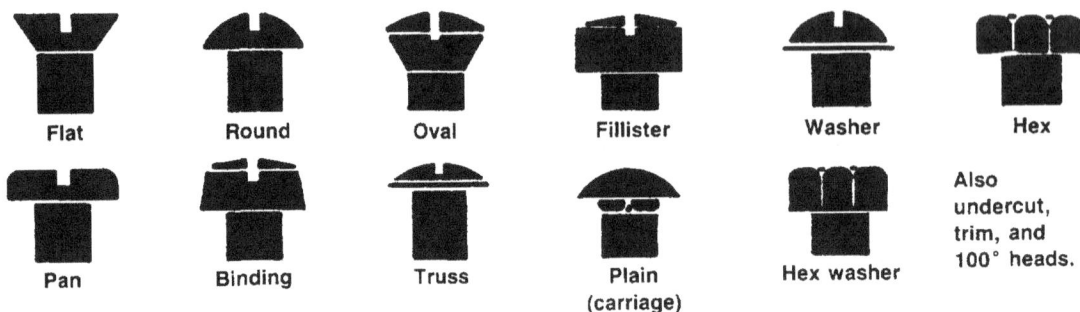

Flat   Round   Oval   Fillister   Washer   Hex

Pan   Binding   Truss   Plain (carriage)   Hex washer

Also undercut, trim, and 100° heads.

Figure 33.—Bolthead and screwhead styles.

Rivets and Lockbolts

Rivets

Rivets are relatively low-cost, permanently installed fasteners that are lighter weight than bolts. As a result, they are the most widely used fasteners in the aircraft manufacturing industry. They are faster to install than bolts and nuts, since they adapt well to automatic, high-speed installation tools. However, rivets should not be used in thick materials or in tensile applications, as their tensile strengths are quite low relative to their shear strengths. The longer the total grip length (the total thickness of sheets being joined), the more difficult it becomes to lock the rivet.

Riveted joints are neither airtight nor watertight unless special seals or coatings are used. Since rivets are permanently installed, they have to be removed by drilling them out, a laborious task.

General Rivet Types

The general types of rivets are solid, blind, tubular, and metal piercing (including split rivets). From a structural design aspect the most important rivets are the solid and blind rivets.

Solid rivets.—Most solid rivets are made of aluminum so that the shop head can be cold formed by bucking it with a pneumatic hammer. Thus, solid rivets must have cold-forming capability without cracking. A representative listing of solid rivets is given in table IX (ref. 21). Some other solid rivet materials are brass, SAE 1006 to SAE 1035, 1108 and 1109 steels, A286 stainless steel, and titanium.

Note that the rivets in table IX are covered by military standard specifications, which are readily available. Although most of the solid rivets listed in table IX have universal heads, there are other common head types, as shown in figure 34. However, because the "experts" do not necessarily agree on the names, other names have been added to the figure. Note also that the countersunk head angle can vary from 60° to 120° although 82° and 100° are the common angles.

TABLE IX.—ALUMINUM AND OTHER RIVET MATERIALS

[From ref. 21.]

Material	Rivet designation	Rivet heads available	Applications
2117–T4	AD	Universal (MS20470) 100° Flush (MS20426)	General use for most applications
2024–T4	DD	Universal (MS20470) 100° Flush (MS20426)	Use only as an alternative to 7050-T73 where higher strength is required
1100	A	Universal (MS20470) 100° Flush (MS20426)	Nonstructural
5056–H32	B	Universal (MS20470) 100° Flush (MS20426)	Joints containing magnesium
Monel (annealed)	M	Universal (MS20615) 100° Flush (MS20427)	Joining stainless steels, titanium, and Inconel
Copper (annealed)	---	100° Flush (MS20427)	Nonstructural
7050–T73	E	Universal (MS20470) 100° Flush (MS20426)	Use only where higher strength is required

The sharp edge of the countersunk head is also removed in some cases, as in the Briles[13] BRFZ "fast" rivet (fig. 35), to increase the shear and fatigue strength while still maintaining a flush fit.

Blind rivets.—Blind rivets get their name from the fact that they can be completely installed from one side. They have the following significant advantages over solid rivets:

(1) Only one operator is required for installation.

(2) The installation tool is portable (comparable to an electric drill in size).

---

[13]Briles Rivet Corporation, Oceanside, California.

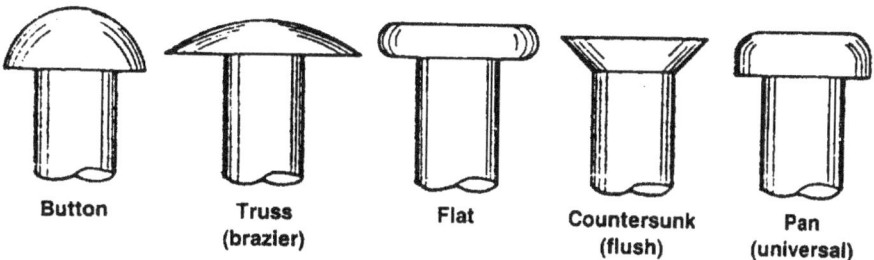

Button **Truss (brazier)** **Flat** **Countersunk (flush)** **Pan (universal)**

Figure 34.—United States standard rivet heads.

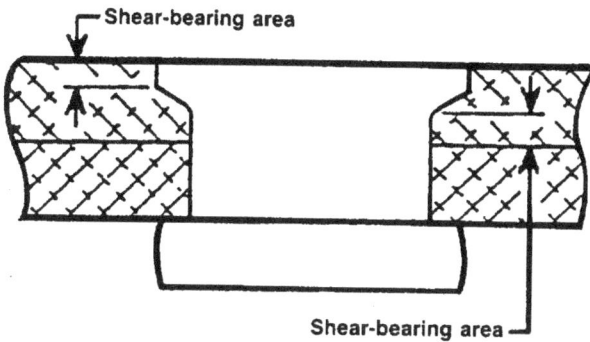

Figure 35.—BRFZ 'fast' rivet.

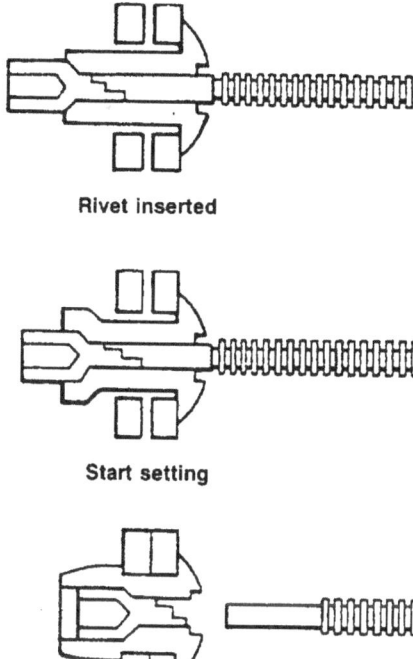

Rivet inserted

Start setting

Figure 36.—Pull-mandrel rivet. (From ref. 5.)

(3) They can be used where only one side of the workpiece is accessible.

(4) A given-length rivet can be used for a range of material thicknesses.

(5) Installation time is faster than with solid rivets.

(6) Clamping force is more uniform than with solid rivets.

(7) Less training is required for the operator.

Blind rivets are classified according to the methods used to install them:

(1) Pull mandrel

(2) Threaded stem

(3) Drive pin

Specific types (brands) of blind rivets are covered in subsequent sections of this manual.

Pull-mandrel rivets: This rivet is installed with a tool that applies force to the rivet head while pulling a prenotched serrated mandrel through to expand the far side of the tubular rivet. When the proper load is reached, the mandrel breaks at the notch. A generic pull-mandrel rivet is shown in figure 36.

Threaded-stem rivets: The threaded-stem rivet (fig. 37(a)) has a threaded internal mandrel (stem) with the external portion machined flat on two sides for the tool to grip and rotate. The head is normally hexagonal to prevent rotation of the tubular body while the mandrel in being torqued and broken off.

Drive-pin rivets: This rivet has a drive pin that spreads the far side of the rivet to form a head, as shown in figure 38. Although drive-pin rivets can be installed quickly, they are

usually not used in aerospace applications. They are used primarily for commercial sheet metal applications.

Tubular rivets.—Tubular rivets are partially hollow and come in a variety of configurations. The generic form has a manufactured head on one side and a hollow end that sticks through the pieces being joined. The hollow end is cold formed to a field head.

Since extensive cold forming is required on these rivets, they must be extremely ductile and are consequently made of low-strength materials. They are normally used for commercial applications rather than in the aerospace industry.

Some specific types of tubular rivets are

(1) Compression

(2) Semitubular

(3) Full tubular

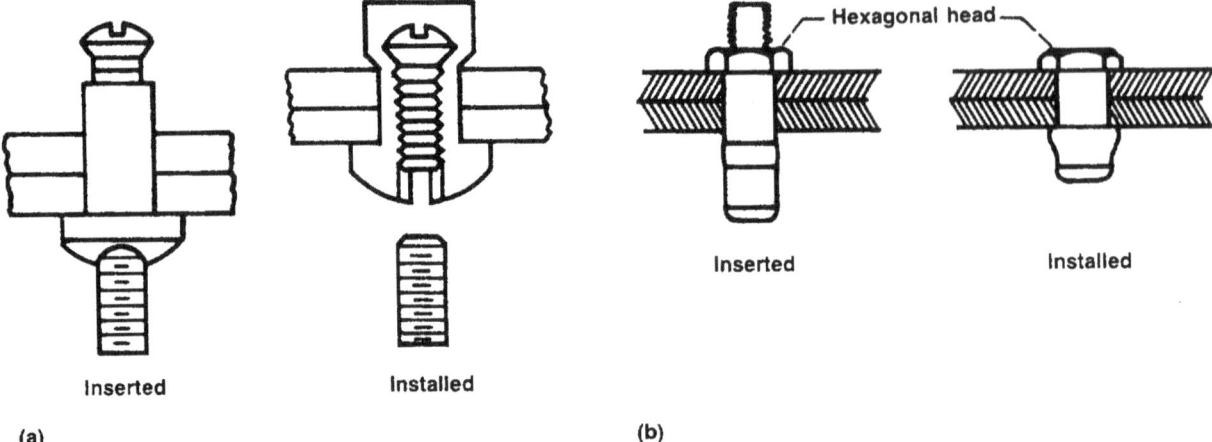

Inserted   Installed   Inserted   Installed

(a)   **(b)**

(a) One-piece body. (From ref. 5.)
(b) Two-piece body. (From ref. 22.)

Figure 37.—Threaded-stem rivets.

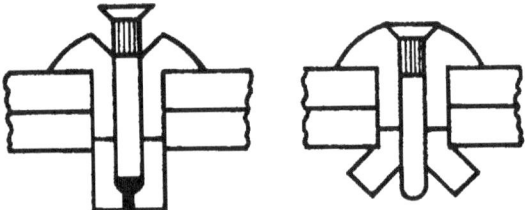

Figure 38.—Drive-pin rivet. (From ref. 5.)

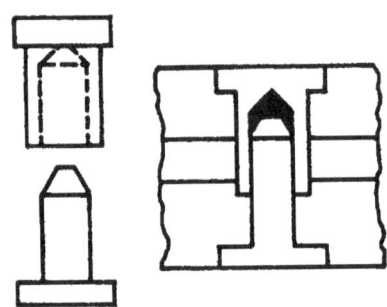

Figure 39.—Compression tubular rivet. (From ref. 5.)

Compression tubular rivets: A compression tubular rivet (fig. 39) consists of two parts that have an interference fit when driven together. These rivets are used commercially in soft materials and where a good appearance is required on both sides of the part.

Semitubular rivets: The semitubular rivet (fig. 40) has a hole in the field end (hole depth to 1.12 of shank diameter) such that the rivet approaches a solid rivet when the field head is formed.

Full tubular rivets: The full tubular rivet (fig. 41) has a deeper hole than the semitubular rivet. It is a weaker rivet than the semitubular rivet, but it can pierce softer materials such as plastic or fabric.

Metal-piercing rivets.—Metal piercing rivets (fig. 42) are similar to semitubular rivets, except that they have greater column strength. Part of the sandwich material is not drilled, and the rivet pierces all the way or most of the way through while mushrooming out to a locked position.

Figure 40.—Semitubular rivet. (From ref. 5.)

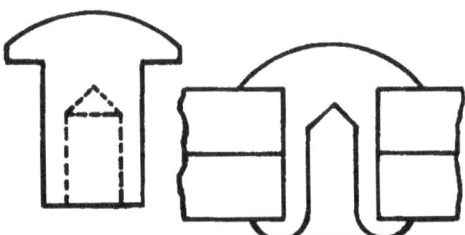

Figure 41.—Full tubular rivet. (From ref. 5.)

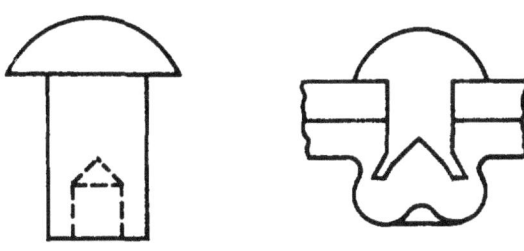

Figure 42.—Metal-piercing rivet. (From ref. 5.)

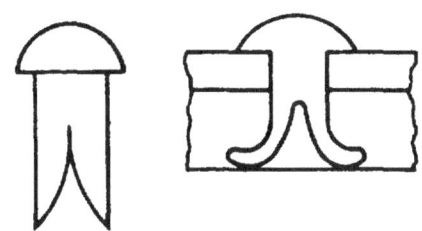

Figure 43.—Split (bifurcated) rivet. (From ref. 5.)

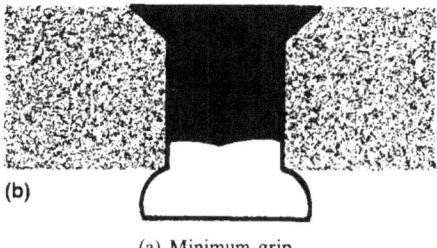

(a) Minimum grip.
(b) Maximum grip.

Figure 44.—Cherry Buck rivet.

Split rivets.—Split (bifurcated) rivets (fig. 43) are the standard "home repair" rivets. They have sawed or split bodies with sharp ends to make their own holes through leather, fiber, plastic, or soft metals. They are not used in critical applications.

Specific Rivet Types

AD & DD solid rivets.—The most common solid rivets are the AD and DD aluminum rivets, as listed in table IX. These are the preferred rivets for joining aluminums and combinations of aluminum and steel. The "icebox" (DD) rivets can be used in higher-strength applications, but they must be kept around 0 °F until they are installed. The 7050–T73 aluminum rivets are an alternative to "icebox" rivets.

Since solid rivets are expanded to an interference fit, they *should not* be used in composites or fiber materials. They can cause delamination of the hole surfaces, leading to material failure.

Cherry Buck rivets.—The Cherry Buck rivet[14] is a hybrid consisting of a factory head and shank of 95-ksi-shear-strength titanium, with a shop end shank of ductile titanium/niobium, joined together by inertia welding (fig. 44). This combination allows a shop head to be formed by bucking, but the overall shear strength of the rivet approaches 95 ksi. The Cherry Buck rivet can be used to 600 °F.

Monel rivets.—Monel (67 percent nickel and 30 percent copper) rivets are used for joining stainless steels, titanium, and Inconel. Monel is ductile enough to form a head without cracking but has higher strength ($F_{su} = 49$ ksi) and temperature capabilities than aluminum.

Titanium/niobium rivets.—These titanium alloy rivets (per MIL–R–5674 and AMS4982) have a shear strength of 50 ksi but are still formable at room temperature. They generally do not need a coating for corrosion protection. The Cherry E–Z Buck is a titanium/niobium rivet.

Cherry rivets.—The generic Cherry rivet is a blind structural rivet with a locking collar for the stem installed as shown in figure 45. (Different head types are available.) Cherry rivets are available in both nominal and oversize diameters in the common (⅛ through ¼ in.) sizes. The oversize rivets are used for repairs where a nominal-size rivet (solid or blind) has been drilled out or where the initial drilled hole is oversize. These rivets have shear strengths comparable to AD solid aluminum rivets. However, their usage is restricted in aircraft manufacturing by the guidelines of MS33522, which is included as appendix C. A typical list of available Cherry rivet materials is shown in table X.

Huck blind rivets.—Huck blind rivets[15] are similar to Cherry rivets, except that they are available in higher strength material. These rivets are made with and without locking collars and with countersunk or protruding heads. Note also (in fig. 46) that the sleeve on the blind side is deformed differently on the Huck rivet than on the Cherry rivet.

[14]Townsend Company, Cherry River Division, Santa Ana, California.

[15]Huck Manufacturing Company, Long Beach, California.

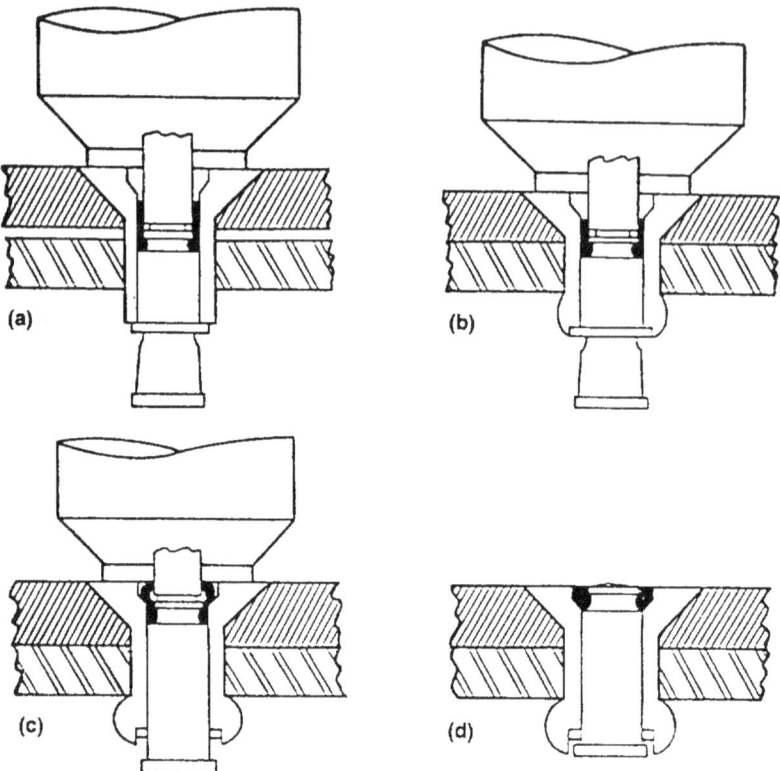

(a) Insert CherryMAX rivet into prepared hole. Place pulling head over rivet stem and apply firm, steady pressure to seat head. Actuate tool.
(b) Stem pulls into rivet sleeve and forms large bulbed blind head; seats rivet head and clamps sheets tightly together. Shank expansion begins.
(c) "Safe-lock" locking collar moves into rivet sleeve recess. Formation of blind head is completed. Shear-ring has sheared from cone, thereby accommodating a minimum of $\frac{1}{16}$ in. in structure thickness variation.
(d) Driving anvil forms "safe-lock" collar into head recess, locking stem and sleeve securely together. Continued pulling fractures stem, providing flush, burr-free, inspectable installation.

Figure 45.—Cherry rivet installation.

TABLE X.—CHERRY RIVET MATERIALS

Materials		Ultimate shear strength, psi	Maximum temperature, °F
Sleeve	Stem		
5055 Aluminum	Alloy steel	50 000	250
5056 Aluminum	CRES	50 000	250
Monel	CRES	55 000	900
Inco 600	Inco X750	75 000	1400

Pop rivets.—Pop rivets[16] are familiar to most of the public for home repairs. However, they are not recommended for critical structural applications. The stem sometimes falls out of the sleeve after the rivet is installed, and the symmetry of the blind (formed) head leaves much to be desired. Although the pop rivet shown in figure 47 is the most common type, USM makes a closed-end rivet and three different head styles.

---

[16]USM Corporation, Pop Rivet Division, Shelton, Connecticut.

Lockbolts

In general, a lockbolt is a nonexpanding, high-strength fastener that has either a swaged collar or a type of threaded collar to lock it in place. It is installed in a standard drilled hole with a snug fit but normally not an interference fit. A lockbolt is similar to an ordinary rivet in that the locking collar or nut is weak in tension loading and is difficult to remove once installed.

Some of the lockbolts are similar to blind rivets and can be completely installed from one side. Others are fed into the workpiece with the manufactured head on the far side. The installation is then completed from the near side with a gun similar to blind rivet guns. Lockbolts are available with either countersunk or protruding heads.

Since it is difficult to determine whether a lockbolt is installed properly, they should be used only where it is not possible to install a bolt and nut of comparable strength. However, they are much faster to install than standard bolts and nuts.

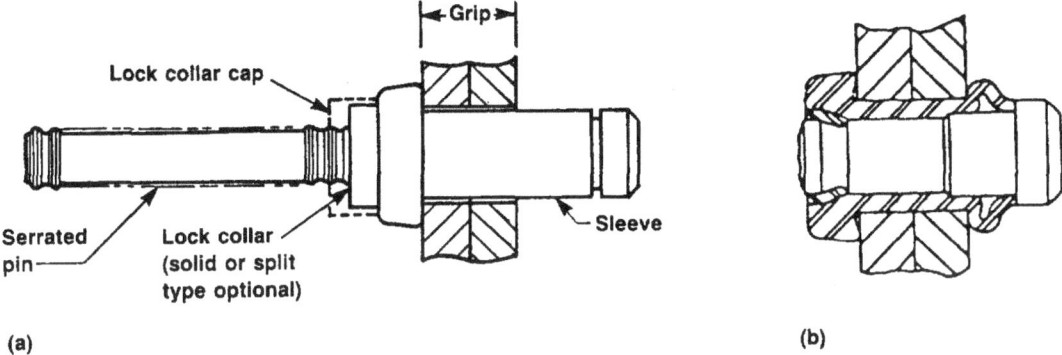

(a) Protruding head, BP-T (MS90354) or BP-EU (MS21141).
(b) Installed fastener.

Figure 46.—Huck blind rivets.

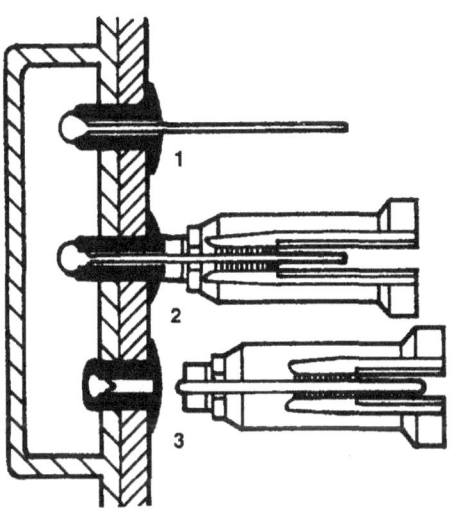

Figure 47.—Pop rivet installation.

Jo-Bolts

Jo-bolts are similar to blind rivets in appearance and installation. The locking collar (sleeve) is expanded to form a shop head by rotating the threaded stem with a gun. The threaded stem is notched and breaks off when the proper torque is reached. A typical Jo-bolt installation is shown in figure 48.

Hi-Lok

The Hi-Lok[17] lockbolt has a countersunk or protruding manufactured head and threads like a bolt. It is fed through the hole from the far side. The installation gun prevents shank rotation with a hexagonal key while the nut is installed (as shown in fig. 49). The nut (collar) hexagonal end is notched to break off at the desired torque. Hi-Lok lockbolts are available in high-strength carbon steel (to 156-ksi shear), stainless steel (to 132-ksi shear), and titanium (to 95-ksi shear).

Huckbolts

Huckbolts[15] are similar to Hi-Loks except that the stem is usually serrated rather than threaded. The collar is swaged on the stem. Then the stem is broken at the notch as shown in figure 50. Huckbolts and their collars are available in carbon steel, aluminum, and stainless steel with various strengths, as listed in the Huck catalog.

Taper-Lok

Taper-Lok[18] is a high-strength threaded fastener that is

---

[17]Hi-Shear Corporation, Torrance, California.
[18]SPS Technologies, Jenkintown, Pennsylvania.

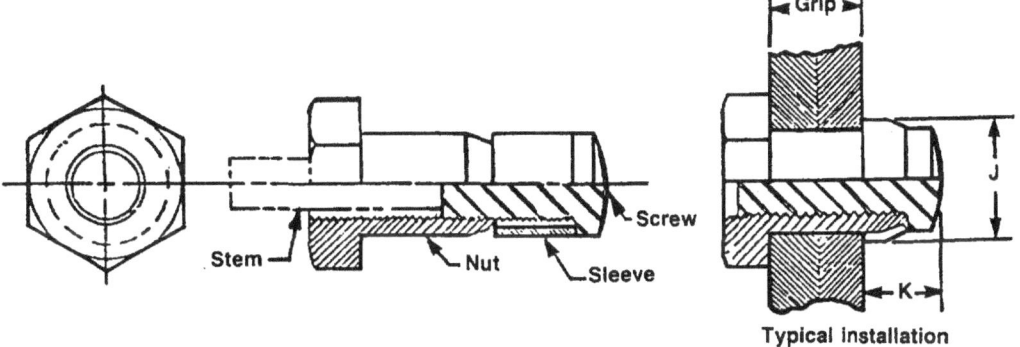

Figure 48.—Jo-bolt. (From ref. 21.)

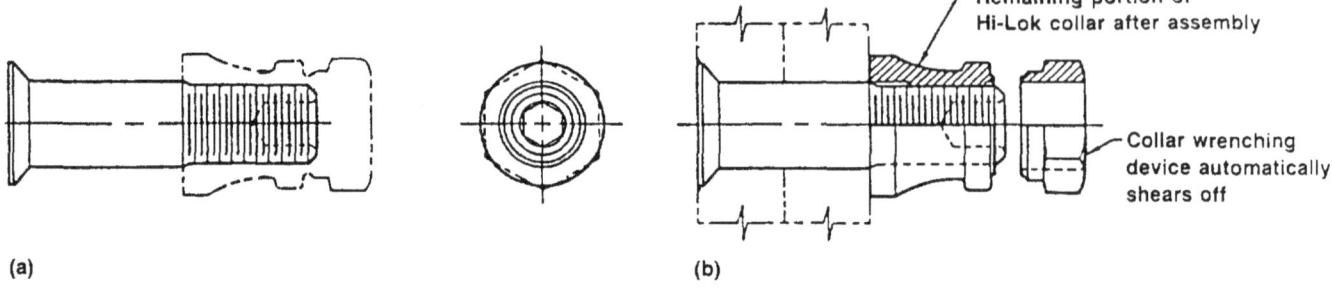

(a) Hi-Lok pin.
(b) Hi-Lok pin and collar after assembly.

Figure 49.—Hi-Lok installation.

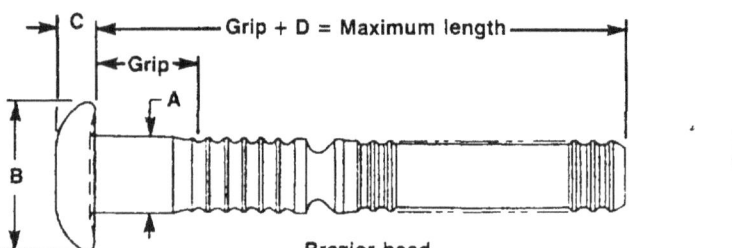 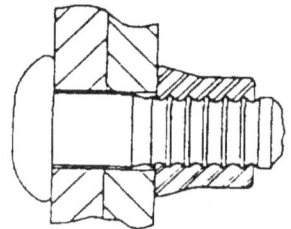

Figure 50.—Installed Huckbolt fastener.

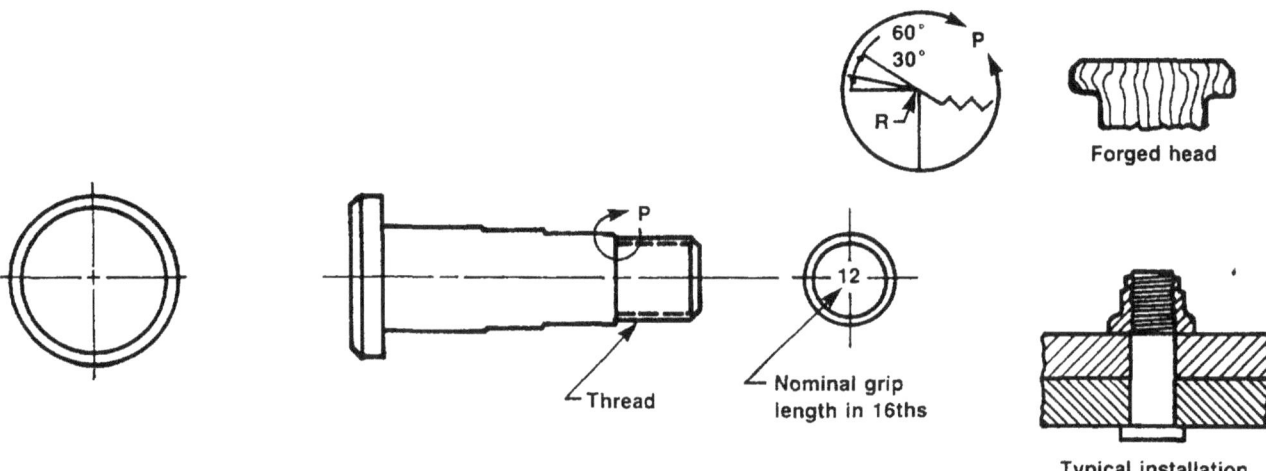

Figure 51.—Taper-Lok installation.

installed with an interference fit. Most of the shank is tapered on a 1.19° angle. The lubricated lockbolt is driven into a drilled and reamed hole. The interference fit allows the nut (tension or shear nut) to be installed and torqued to the required value without holding the lockbolt to prevent rotation (see fig. 51). The nuts are locknuts with captive washers. When a tension nut is installed, this fastener can take as much tension load as a bolt of the same size and material. Consequently, Taper-Loks are used in critical applications where cyclic loading is a problem. Taper-Lok lockbolts are available in high-strength alloy steel, H-11 tool steel, and several stainless steels, as well as titanium.

Rivnuts

A Rivnut[19] is a tubular rivet with internal threads that is deformed in place to become a blind nutplate (fig. 52). Rivnuts are available with protruding, countersunk, and fillister heads. They are also available with closed ends, sealed heads, ribbed shanks, hexagonal shanks, and ribbed heads. Since the unthreaded tubular portion of the rivet must deform, the material must be ductile. Consequently, the Rivnut materials are fairly low strength, as shown in table XI.

---

[19]B.F. Goodrich, Engineered Systems Division. Akron, Ohio.

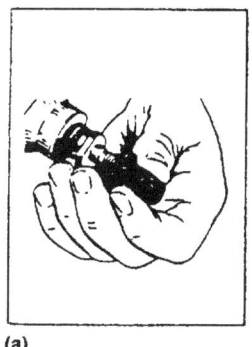 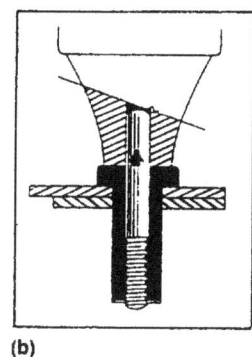 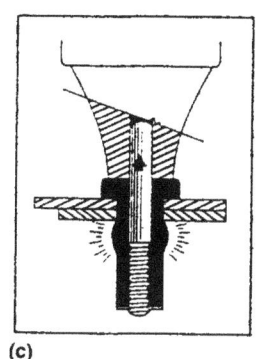 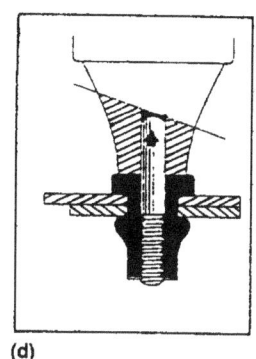 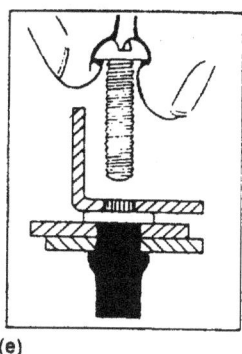

(a) (b) (c) (d) (e)

(a) Step 1—Rivnut fastener is threaded onto mandrel of installation tool.

(b) Step 2—Rivnut fastener, on tool mandrel, is inserted into hole drilled for installation.

(c) Step 3—Mandrel retracts and pulls threaded portion of Rivnut fastener shank toward blind side of work, forming bulge in unthreaded shank area.

(d) Step 4—Rivnut fastener is clinched securely in place; mandrel is unthreaded, leaving internal Rivnut threads intact.

(e) Blind nutplate—Properly installed Rivnut fastener makes excellent blind nutplate for simple screw attachments; countersunk Rivnut fasteners can be used for smooth surface installation.

Figure 52.—Rivnut installation.

TABLE XI.—STANDARD RIVNUT FASTENER
MATERIALS AND FINISHES

Material	Type	Standard finish	Minimum ultimate tensile strength, psi
Aluminum	6053–T4	Anodize—Alumilite 205 will meet specifications: MIL-A-8625 (ASG)	28 000
Steel	C–1108[a] C–1110[a]	Cadmium plate—0.0002 in. minimum thickness per QQ-P-416b, class 3, type I	45 000
	4037	Cadmium plate—0.0002 in. minimum thickness per QQ-P-416b, class 2, type II	[b]55 000 [c]85 000
Stainless steel	430	Pickled and passivated per QQ-P-35, type II	67 000
	305[d] Carpenter 10[d]	None—bright as machined	80 000
Brass	Alloy 260	None—bright as machined	50 000

[a]C-1108 and C-1110 steel may be used interchangeably.
[b]No. 4 and No. 6 thread sizes.
[c]No. 8—1/2-in. thread size.
[d]305 and Carpenter No. 10 stainless steel may be used interchangeably.

Hi-Shear Rivet

Hi-Shear[17] rivets consist of a high-strength carbon steel, stainless steel, aluminum, or titanium rivet (pin) with a necked-down shop head, as shown in figure 53. The collar (2024 aluminum or Monel) is swaged on to give a finished head that

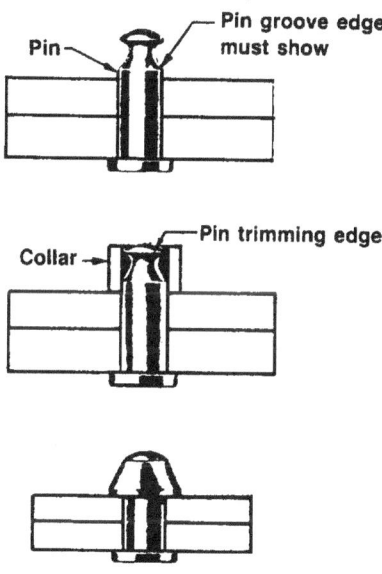

Figure 53.—Hi-Shear installation.

can be visually inspected for proper form. This rivet should be used for shear applications only, as the collar has negligible tensile strength.

Although this rivet has been partially superseded by various lockbolts, it is still being used in aircraft and aerospace applications.

Lightweight Grooved Proportioned Lockbolt

The lightweight grooved proportioned lockbolt (LGPL)[20] is made especially for composite materials. It has both an oversize head and an oversize collar to lessen contact stresses

[20]Monogram Aerospace Fasteners, Los Angeles, California.

33

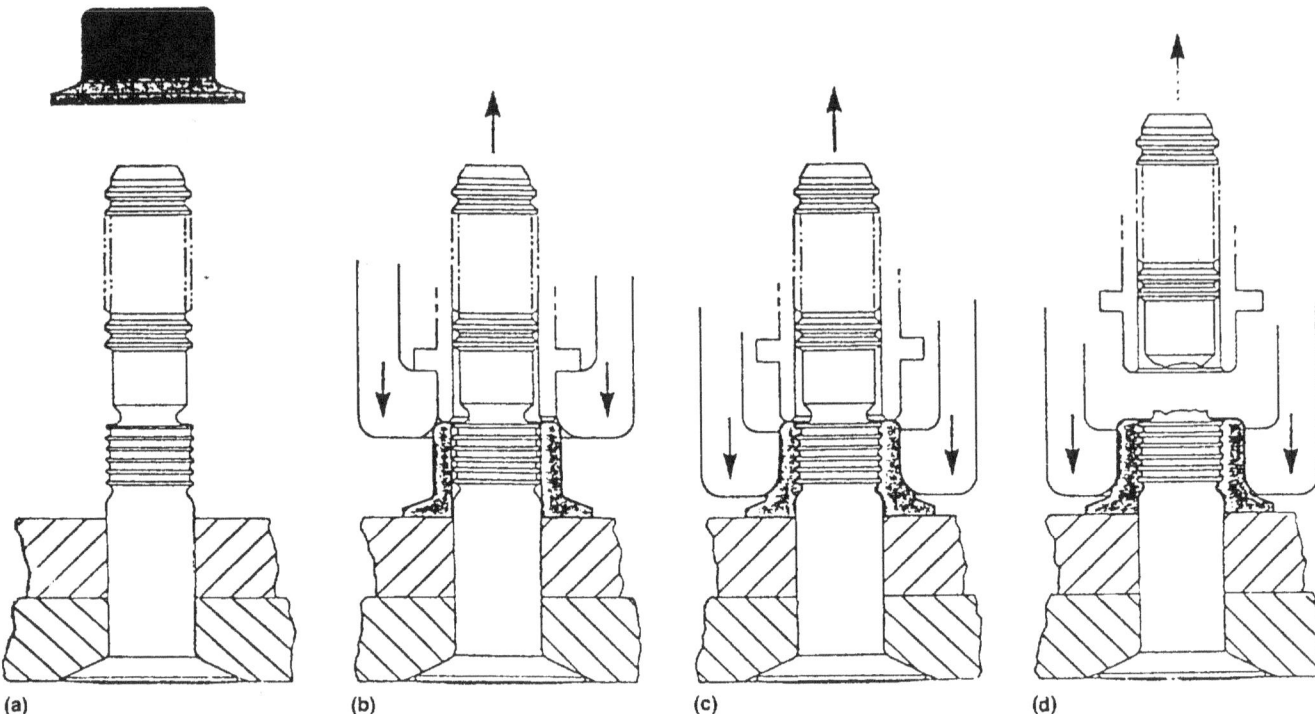

(a) Flanged collar is placed over lightweight pin.
(b) Installation tool grips and pulls pin, drawing sheets tightly together and removing sheet gap.
(c) As pull on pin increases, tool anvil swages flanged collar into locking grooves and forms permanent vibration-resistant lock.
(d) Pull on pin continues until pin fractures at breakneck groove and is then ejected. Tool anvil disengages swaged collar.

Figure 54.—LGPL installation.

on the composite material during both installation and service life. The shank is high-strength (95-ksi shear) titanium and the collar is 2024 aluminum. It is installed with a lockbolt tool as shown in figure 54.

General Guidelines for Selecting Rivets and Lockbolts

A number of standard documents are available for the selection, installation, and drawing callout of rivets and lockbolts as follows:

(1) Rivet installations are covered by MIL-STD-403. This specification covers pilot holes, deburring, countersinking, dimpling, and the application of zinc chromate paint between dissimilar materials. Other specifications for corrosion prevention of drilled or countersunk surfaces are covered in MIL-P-116 and MIL-STD-171.

(2) Design and selection requirements for blind *structural* rivets are given in MS33522 (appendix C).

(3) Design and selection requirements for blind *nonstructural* rivets are given in MS33557.

(4) A wealth of information on allowable rivet strengths in various materials and thicknesses is given in chapter 8 of MIL-HDBK-5 (ref. 18).

(5) Testing of fasteners is covered by MIL-STD-1312.

(6) Lockwiring is done per MS33540.

Note that the nominal rivet spacing for a rivet pattern is an edge distance of $2D$ and a linear spacing of $4D$, where D is the rivet diameter. However, the $4D$ spacing can be increased if sealing between rivets or interrivet buckling is not a problem.

Solid rivets (expanded during installation) should not be used in composite materials, as they can overstress the hole and cause delamination of the material.

Lewis Research Center
National Aeronautics and Space Administration
Cleveland, Ohio, June 30, 1989

References

1. Sliney, H.E.: High Temperature Solid Lubricants—1. Layer Lattice Compounds and Graphite. Mech. Eng., vol. 96, no. 2, Feb. 1974, pp. 18-22.

2. Prevention of Material Deterioration: Corrosion Control Course—U.S. Army Logistics Engineering Directorate—Nov. 1970.

3. ASM Metals Handbook. 9th ed., Vols. 1, 2, 3, 5, 13, American Society for Metals, Metals Park, OH.

4. SAE Handbook. SAE, 1968.

5. 1987 Fastening, Joining & Assembly Reference Issue. Mach. Des., vol. 59, no. 27, Nov. 19, 1987.

6. Unified Inch Screw Threads (UN and UNR Thread Form). ANSI B1.1-1982, American National Standards Institute, New York, NY, 1982.

7. Screw Thread Standards for Federal Services, Part 1—Unified UNJ Unified Miniature Screw Threads. National Bureau of Standards Handbook, NBS-H28-1969-PT-1, 1969.

8. Fastener Standards. 5th ed., Industrial Fasteners Institute, Cleveland, OH, 1970.

9. Bickford, J.H.: An Introduction to the Design and Behavior of Bolted Joints. Dekker, 1981.

10. Juvinall, R.: Engineering Considerations of Stress, Strain, and Strength. McGraw-Hill, 1967.

11. Donald, E.P.: A Practical Guide to Bolt Analysis. Mach. Des., vol. 53, Apr. 9, 1981, pp. 225-231.

12. Baumeister, et al.: Mark's Standard Handbook for Mechanical Engineers. 8th ed., McGraw-Hill, 1978.

13. Seely, F.B.: Resistance of Materials. 3rd ed., Wiley & Sons, 1947.

14. Shigley, J.E.; and Mitchell, L.D.: Mechanical Engineering Design. 4th ed., McGraw-Hill, 1983.

15. Machine Design, Nov. 19, 1981.

16. Peery, D.J.: Aircraft Structures. McGraw-Hill, 1950.

17. Grinter, L.: Theory of Modern Steel Structures. Vol. I, Macmillan Co., 1955.

18. Metallic Materials and Elements for Aerospace Vehicle Structures. MIL-HDBK-5E, Department of Defense, June 1987.

19. Faupel, J.H.; and Fisher, F.E.: Engineering Design, 2nd ed., Wiley & Sons, 1981.

20. Fastener Technology International Magazine, Solon, Ohio, Oct. 1985 through Feb. 1987 Editions.

21. Design Handbook, Section 16. McDonnell Douglas Astronautics Co., Huntington Beach, CA.

22. Bruhn, E.F.: Analysis & Design of Flight Vehicle Structures. Tri-State Offset Co., Cincinnati, 1965.

Appendix A
Bolthead Marking and Design Data

[From ref. 20]

ASTM markings

The American Society for Testing and Materials, 1916 Race St. Philadelphia, PA 19103, sponsors development of specifications for fasteners used in general and special engineering applications. These specifications detail chemical and mechanical properties of material strength levels for fasteners and are generally specific in referencing the actual product covered. A full range of types of products of various styles, thread series, lengths, etc, can be produced to meet ASTM requirements and would be marked for grade and material identification as required.

ASTM Grade and material identification markings required by ASTM specifications

Grade ID mark	ASTM spec number	Fastener description	Material	Is mfgr's ID symbol req'd?	Nominal size range (inch)	Bolts, screws, studs Proof load (psi)	Bolts, screws, studs Yield strength (min psi)	Bolts, screws, studs Tensile strength (min psi)	Nuts Proof load (psi)	Hardness Brinell	Hardness Rockwell	Remarks or footnotes
None req'd	A31, Grade A	Boiler rivets	Carbon steel	No	Thru 1½	—	23,000	45,000	—	—	—	—
None req'd	A31, Grade B	Boiler rivets	Carbon steel	No	Thru 1½	—	29,000	58,000	—	—	—	—
None req'd	A65, Grade 1	Track spikes	Carbon steel, copper not specified	Yes	—	—	0.5X tensile strength	55,000	—	—	—	Marking on top of head.
CU	A65, Grade 1	Track spikes	Carbon steel, copper specified	Yes	—	—	0.5X tensile strength	55,000	—	—	—	Marking on top of head.
HC	A65, Grade 2	Track spikes	Carbon steel, copper not specified	Yes	—	—	0.5X tensile strength	70,000	—	—	—	Marking on top of head
HC and CU	A65, Grade 2	Track spikes	Carbon steel, copper specified	Yes	—	—	0.5X tensile strength	70,000	—	—	—	Marking on top of head
None req'd	A66	Screw spikes	Carbon steel	Yes	—	—	0.5X tensile strength	60,000	—	—	—	Marking on spike head.
None req'd	A183, Grade 1	Track bolts	Low carbon steel, untreated	Yes	½ - 1⅛	—	—	55,000	—	—	—	Marking on top of head, raised or depressed
See "Remarks"	A183, Grade 2	Track bolts	Carbon steel, heat-treated	Yes	½ - 1⅛	—	80,000	110,000	—	—	—	Marking on top of head, raised or depressed A symbol is required to indicate bolt is heat-treated.
None req'd	A183, Grade 1	Track nuts	Low carbon steel	No	¾ - 1⅛	—	—	—	—	—	—	—
None req'd	A183, Grade 2	Track nuts	Medium carbon steel	No	¾ - 1⅛	—	—	—	—	—	—	—
B5	A193	Bolts, screws, and studs for high temperature service	AISI 501	Yes	¼ - 4	—	80,000	100,000	—	—	—	A
B6	A193	Bolts, screws, and studs for high temperature service	AISI 410	Yes	¼ - 4	—	85,000	110,000	—	—	—	A

Footnotes are grouped on the last page of this Part 1 series.

Footnotes are grouped on the last page of this Part 1 series.

Grade ID mark	ASTM spec number	Fastener description	Is mfgr's ID symbol req'd?	Nominal size range (inch)	Mechanical properties						Remarks or footnotes
					Bolts, screws, studs			Nuts	Hardness		
					Proof load (psi)	Yield strength (min psi)	Tensile strength (min psi)	Proof load (psi)	Brinell	Rockwell	
B6X	A193	Bolts, screws, and studs for high temperature service	Yes	¼ - 4	—	70,000	90,000	—	—	C26 max	A
B7	A193	Bolts, screws and studs for high temperature service	Yes	¼ - 2½ Over 2½ - 4 Over 4 - 7	— — —	105,000 95,000 75,000	125,000 115,000 100,000	— — —	— — —	— — —	A A A
B7M	A193	Bolts, screws and studs for high temperature service	Yes	¼ - 2½	—	80,000	100,000	—	235 max (B)	B99 max (B)	A
B16	A193	Chromium, molybdenum and vanadium alloy steel	Yes	¼ - 2½ Over 2½ - 4 Over 4 - 7	— — —	105,000 95,000 85,000	125,000 110,000 100,000	— — —	— — —	— — —	A A A
B8	A193, A320	Bolts, screws, and studs for high or low temperature service. respectively—Class 1 (C)	Yes	¼ and larger	—	30,000	75,000	—	223[D] max	B96[D] max	A
B8C	A193, A320	Bolts, screws and studs for high or low temperature service. respectively—Class 1 (C)	Yes	¼ and larger	—	30,000	75,000	—	223[D] max	B96[D] max	A
B8M	A193, A320	Bolts, screws and studs for high or low temperature service. respectively—Class 1 (C)	Yes	¼ and larger	—	30,000	75,000	—	223[D] max	B96[D] max	A
B8P	A193, A320	Bolts, screws and studs for high or low temperature service. respectively—Class 1 (C)	Yes	¼ and larger	—	30,000	75,000	—	223[D] max	B96[D] max	A
B8T	A193, A320	Bolts, screws, and studs for high or low temperature service. respectively—Class 1 (C)	Yes	¼ and larger	—	30,000	75,000	—	223[D] max	B96[D] max	A
B8LN	A193, A320	Bolts, screws and studs for high or low temperature service. respectively—Class 1 (C)	Yes	¼ and larger	—	30,000	75,000	—	223[D] max	B96[D] max	A
B8MLN	A193, A320	Bolts, screws, and studs for high or low temperature service. respectively—Class 1 (C)	Yes	¼ and larger	—	30,000	75,000	—	223[D] max	B96[D] max	A

Grade ID mark	ASTM spec number	Fastener description	Material	Is mfgr's ID symbol req'd?	Nominal size range (inch)	Mechanical properties						Remarks or footnotes
						Bolts, screws, studs			Nuts	Hardness		
						Proof load (psi)	Yield strength (min psi)	Tensile strength (min psi)	Proof load (psi)	Brinell	Rockwell	
B8A	A193, A320	Bolts, screws, and studs for high or low temperature service respectively—Class 1A (C)	AISI 304 carbide solution treated in finished condition	Yes	¼ and larger	—	30,000	75,000	—	192 max	B90 max	A
B8CA	A193, A320	Bolts, screws, and studs for high or low temperature service respectively—Class 1A (C)	AISI 347 carbide solution treated in finished condition	Yes	¼ and larger	—	30,000	75,000	—	192 max	B90 max	A
B8MA	A193, A320	Bolts, screws, and studs for high or low temperature service respectively—Class 1A (C)	AISI 316 carbide solution treated in finished condition	Yes	¼ and larger	—	30,000	75,000	—	192 max	B90 max	A
B8PA	A193, A320	Bolts, screws, and studs for high or low temperature service respectively—Class 1A (C)	AISI 305 (with restricted carbon) carbide solution treated in finished condition	Yes	¼ and larger	—	30,000	75,000	—	192 max	B90 max	A
B8TA	A193, A320	Bolts, screws, and studs for high or low temperature service respectively—Class 1A (C)	AISI 321 carbide solution treated in finished condition	Yes	¼ and larger	—	30,000	75,000	—	192 max	B90 max	A
B8LNA	A193, A320	Bolts, screws, and studs for high or low temperature service respectively—Class 1A (C)	AISI 304N (with restricted carbon) carbide solution treated in finished condition	Yes	¼ and larger	—	30,000	75,000	—	192 max	B90 max	A
B8MLNA	A193, A320	Bolts, screws, and studs for high or low temperature service respectively—Class 1A (C)	AISI 316N (with restricted carbon) carbide solution treated in finished condition	Yes	¼ and larger	—	30,000	75,000	—	192 max	B90 max	A
B8NA	A193	Bolts, screws, and studs for high temperature service, Class 1A	AISI 304N carbide solution treated in finished condition	Yes	¼ and larger	—	30,000	75,000	—	192 max	B90 max	A
B8MNA	A193	Bolts, screws, and studs for high temperature service, Class 1A	AISI 316N carbide solution treated in finished condition	Yes	¼ and larger	—	30,000	75,000	—	192 max	B90 max	A
B8N	A193	Bolts, screws, and studs for high temperature service, Class 1B	AISI 304N carbide solution treated	Yes	¼ and larger	—	35,000	80,000	—	223 max (1)	B90 max (1)	A

Footnotes are grouped on the last page of this Part 1 series.

Grade ID mark	ASTM spec number	Fastener description	Material	Is mfgr's ID symbol req'd?	Nominal size range (inch)	Mechanical properties: Bolts, screws, studs: Proof load (psi)	Bolts, screws, studs: Yield strength (min psi)	Bolts, screws, studs: Tensile strength (min psi)	Nuts: Proof load (psi)	Hardness: Brinell	Hardness: Rockwell	Remarks or footnotes
B8MN	A193	Bolts, screws, and studs for high temperature service, Class 1B	AISI 316N, carbide solution treated	Yes	¼ and larger	—	35,000	80,000	—	223 max (D)	B96 max (D)	A
B8R	A193	Bolts, screws, and studs for high temperature service, Class 1C	UNS 20910 (XM19), carbide solution treated	Yes	¼ and larger	—	55,000	100,000	—	271 max	C28 max	A
B8RA	A193	Bolts, screws, and studs for high temperature service, Class 1C	UNS 20910 (XM19), carbide solution treated in finished condition	Yes	¼ and larger	—	55,000	100,000	—	271 max	C28 max	A
B8S	A193	Bolts, screws, and studs for high temperature service, Class 1C	S21800, carbide solution treated	Yes	¼ and larger	—	50,000	95,000	—	271 max	C28 max	A
B8SA	A193	Bolts, screws, and studs for high temperature service, Class 1C	S21800, carbide solution treated in finished condition	Yes	¼ and larger	—	50,000	95,000	—	271 max	C28 max	A
B8	A193, A320	Bolts, screws, and studs for high or low temperature service, respectively—Class 2 (C)	AISI 304, carbide solution treated & strain hardened	Yes	¼ - ¾	—	100,000	125,000	—	321 max	C35 max	A
					Over ¾ - 1	—	80,000	115,000	—	321 max	C35 max	A
					Over 1 - 1¼	—	65,000	105,000	—	321 max	C35 max	A
					Over 1¼ - 1½	—	50,000	100,000	—	321 max	C35 max	A
B8C	A193, A320	Bolts, screws, and studs for high or low temperature service, respectively—Class 2 (C)	AISI 347 carbide solution treated & strain hardened	Yes	¼ - ¾	—	100,000	125,000	—	321 max	C35 max	A
					Over ¾ - 1	—	80,000	115,000	—	321 max	C35 max	A
					Over 1 - 1¼	—	65,000	105,000	—	321 max	C35 max	A
					Over 1¼ - 1½	—	50,000	100,000	—	321 max	C35 max	A
B8P	A193, A320	Bolts, screws, and studs for high or low temperature service, respectively—Class 2 (C)	AISI 305 (with restricted carbon), carbide solution treated & strain hardened	Yes	¼ - ¾	—	100,000	125,000	—	321 max	C35 max	A
					Over ¾ - 1	—	80,000	115,000	—	321 max	C35 max	A
					Over 1 - 1¼	—	65,000	105,000	—	321 max	C35 max	A
					Over 1¼ - 1½	—	50,000	100,000	—	321 max	C35 max	A
B8T	A193, A320	Bolts, screws, and studs for high or low temperature service, respectively—Class 2 (C)	AISI 321, carbide solution treated & strain hardened	Yes	¼ - ¾	—	100,000	125,000	—	321 max	C35 max	A
					Over ¾ - 1	—	80,000	115,000	—	321 max	C35 max	A
					Over 1 - 1¼	—	65,000	105,000	—	321 max	C35 max	A
					Over 1¼ - 1½	—	50,000	100,000	—	321 max	C35 max	A

Grade ID mark	ASTM spec number	Fastener description	Material	Is mfgr's ID symbol req'd?	Nominal size range (inch)	Bolts, screws, studs			Nuts	Hardness		Remarks or footnotes
						Proof load (psi)	Yield strength (min psi)	Tensile strength (min psi)	Proof load (psi) hvy hex / hex	Brinell	Rockwell	
B8N	A193	Bolts, screws, and studs for high temperature service. Class 2	AISI 304N, carbide solution treated & strain hardened	Yes	¼ - ¾	—	100,000	125,000	—	321 max	C35 max	A
					Over ¾ - 1	—	80,000	115,000	—	321 max	C35 max	A
					Over 1 - 1¼	—	65,000	105,000	—	321 max	C35 max	A
					Over 1¼ - 1½	—	50,000	100,000	—	321 max	C35 max	A
B8M	A193, A320	Bolts, screws, and studs for high or low temperature service. respectively—Class 2 (C)	AISI 316, carbide solution treated & strain hardened	Yes	¼ - ¾	—	95,000	110,000	—	321 max	C35 max	A
					Over ¾ - 1	—	80,000	100,000	—	321 max	C35 max	A
					Over 1 - 1¼	—	65,000	95,000	—	321 max	C35 max	A
					Over 1¼ - 1½	—	50,000	90,000	—	321 max	C35 max	A
B8MN	A193	Bolts, screws, and studs for high temperature service. Class 2	AISI 316N, carbide solution treated & strain hardened	Yes	¼ - ¾	—	95,000	110,000	—	321 max	C35 max	A
					Over ¾ - 1	—	80,000	100,000	—	321 max	C35 max	A
					Over 1 - 1¼	—	65,000	95,000	—	321 max	C35 max	A
					Over 1¼ - 1½	—	50,000	90,000	—	321 max	C35 max	A
1	A194	Hot or cold forged nuts for high pressure & high temperature service	Carbon steel	Yes	¼ and larger	—	—	—	130,000 / 120,000	121 min	B70 min	—
1B	A194	Nuts machined from bars for high pressure & high temperature service	Carbon steel	Yes	¼ and larger	—	—	—	130,000 / 120,000	121 min	B70 min	—
2	A194	Hot or cold forged nuts for high pressure & high temperature service	Carbon steel	Yes	¼ and larger	—	—	—	150,000 / 135,000	159/352	B84 min	—
2B	A194	Nuts machined from bars for high pressure & high temperature service	Carbon steel	Yes	¼ and larger	—	—	—	150,000 / 135,000	159/352	B84 min	—
2H	A194	Hot or cold forged nuts for high pressure & high temperature service	Carbon steel, heat treated	Yes	¼ and larger	—	—	—	175,000 / 150,000	248/352	C24/C38	—
2HB	A194	Nuts machined from bars for high pressure & high temperature service	Carbon steel, heat treated	Yes	¼ and larger	—	—	—	175,000 / 150,000	248/352	C24/C38	E
2HM	A194	Hot or cold forged nuts for high pressure & high temperature service	Carbon steel, heat treated	Yes	¼ and larger	—	—	—	150,000 / 135,000	159/237	C22 max	—

Grade ID mark	ASTM spec number	Fastener description	Material	Is mfgr's ID symbol req'd?	Nominal size range (inch)	Bolts, screws, studs			Nuts	Hardness		Remarks or footnotes
						Proof load (psi)	Yield strength (min psi)	Tensile strength (min psi)	Proof load (psi) hvy hex / hex	Brinell	Rockwell	
2HMB	A194	Nuts machined from bars for high pressure & high temperature service	Carbon steel, heat treated	Yes	¼ and larger	—	—	—	150,000 / 135,000	159/237	C22 max	E
3	A194	Hot or cold forged nuts for high pressure & high temperature service	AISI 501, heat treated	Yes	¼ and larger	—	—	—	175,000 / 150,000	248/352	C24/C38	—
3B	A194	Nuts machined from bars for high pressure & high temperature service	AISI 501, heat treated	Yes	¼ and larger	—	—	—	175,000 / 150,000	248/352	C24/C38	E
4	A194	Hot or cold forged nuts for high pressure & high temperature service	Carbon, molybdenum, heat treated	Yes	¼ and larger	—	—	—	175,000 / 150,000	248/352	C24/C38	—
4B	A194	Nuts machined from bars for high pressure & high temperature service	Carbon, molybdenum, heat treated	Yes	¼ and larger	—	—	—	175,000 / 150,000	248/352	C24/C38	E
6	A194	Hot or cold forged nuts for high pressure & high temperature service	AISI 410, heat treated	Yes	¼ and larger	—	—	—	150,000 / 135,000	228/271	C20/C28	—
6B	A194	Nuts machined from bars for high pressure & high temperature service	AISI 410, heat treated	Yes	¼ and larger	—	—	—	150,000 / 135,000	228/271	C20/C28	E
6F	A194	Hot or cold forged nuts for high pressure & high temperature service	AISI 416 with sulfur or 416Se with selenium, heat treated	Yes	¼ and larger	—	—	—	150,000 / 135,000	228/271	C20/C28	—
6FB	A194	Nuts machined from bars for high pressure & high temperature service	AISI 416 with sulfur or 416Se with selenium, heat treated	Yes	¼ and larger	—	—	—	150,000 / 135,000	228/271	C20/C28	E
7	A194	Hot or cold forged nuts for high pressure & high temperature service	AISI 4140/4142/4145, 4140H, 4142H, 4145H, heat treated	Yes	¼ and larger	—	—	—	175,000 / 150,000	248/352	C24/C38	—
7B	A194	Nuts machined from bars for high pressure & high temperature service	AISI 4140/4142/4145, 4140H, 4142H, 4145H, heat treated	Yes	¼ and larger	—	—	—	175,000 / 150,000	248/352	C24/C38	E
7M	A194	Hot or cold forged nuts for high pressure & high temperature service	AISI 4140/4142/4145, 4140H, 4142H, heat treated	Yes	¼ and larger	—	—	—	150,000 / 135,000	159/237	C22 max	—
7MB	A194	Nuts machined from bars for high pressure & high temperature service	AISI 4140/4142/4145, 4140H, 4142H, 4145H, heat treated	Yes	¼ and larger	—	—	—	150,000 / 135,000	159/237	C22 max	E
8	A194	Hot or cold forged nuts for high pressure & high temperature service	AISI 304	Yes	¼ and larger	—	—	—	80,000 / 75,000	126/300	B60/B105	—

Footnotes are grouped on the last page of this Part 1 series.

Grade, ID mark	ASTM spec number	Fastener description	Material	Is mfgr's ID symbol req'd?	Nominal size range (inch)	Mechanical properties						Remarks or footnotes
						Bolts, screws, studs			Nuts	Hardness		
						Proof load (psi)	Yield strength (min psi)	Tensile strength (min psi)	Proof load (psi) hvy hex / hex	Brinell	Rockwell	
8B	A194	Nuts machined from bars for high pressure & high temperature service	AISI 304	Yes	¼ and larger	—	—	—	80,000/75,000	126/300	B60/B105	—
8A	A194	Hot or cold forged or machined from bars for high pressure & high temperature service	AISI 304, carbide solution treated	Yes	¼ and larger	—	—	—	80,000/75,000	126/192	B60/B90	—
8C	A194	Hot or cold forged nuts, for high pressure & high temperature service	AISI 347	Yes	¼ and larger	—	—	—	80,000/75,000	126/300	B60/B105	—
8CB	A194	Nuts machined from bars for high pressure & high temperature service	AISI 347	Yes	¼ and larger	—	—	—	80,000/75,000	126/300	B60/B105	—
8CA	A194	Hot or cold forged or machined from bars for high pressure & high temperature service	AISI 347, carbide solution treated	Yes	¼ and larger	—	—	—	80,000/75,000	126/192	B60/B90	—
8M	A194	Hot or cold forged nuts for high pressure & high temperature service	AISI 316	Yes	¼ and larger	—	—	—	80,000/75,000	126/300	B60/B105	—
8MB	A194	Nuts machined from bars for high pressure & high temperature service	AISI 316	Yes	¼ and larger	—	—	—	80,000/75,000	126/300	B60/B105	—
8MA	A194	Hot or cold forged or machined from bars for high pressure & high temperature service	AISI 316, carbide solution treated	Yes	¼ and larger	—	—	—	80,000/75,000	126/192	B60/B90	—
8T	A194	Hot or cold forged nuts for high pressure & high temperature service	AISI 321	Yes	¼ and larger	—	—	—	80,000/75,000	126/300	B60/B105	—
8TB	A194	Nuts machined from bars for high pressure & high temperature service	AISI 321	Yes	¼ and larger	—	—	—	80,000/75,000	126/300	B60/B105	—
8TA	A194	Hot or cold forged or machined from bars for high pressure & high temperature service	AISI 321, carbide solution treated	Yes	¼ and larger	—	—	—	80,000/75,000	126/192	B60/B90	—
8F	A194	Hot or cold forged nuts for high pressure & high temperature service	AISI 303 with sulfur or 303Se with selenium	Yes	¼ and larger	—	—	—	80,000/75,000	126/300	B60/B105	—
8FB	A194	Nuts machined from bars for high pressure & high temperature service	AISI 303 with sulfur or 303Se with selenium	Yes	¼ and larger	—	—	—	80,000/75,000	126/300	B60/B105	—

Footnotes are grouped on the last page of this Part 1 series.

43

Grade ID mark	ASTM spec number	Fastener description	Material	Is mfgr's ID symbol req'd?	Nominal size range (inch)	Bolts, screws, studs			Nuts	Hardness		Remarks or footnotes
						Proof load (psi)	Yield strength (min psi)	Tensile strength (min psi)	Proof load (psi) hvy hex / hex	Brinell	Rockwell	
8FA	A194	Hot or cold forged or machined from bars for high pressure & high temperature service	AISI 303 with sulfur or 303Se with selenium, carbide solution treated	Yes	⅛ and larger	—	—	—	80,000 / 75,000	126/192	B60/B90	—
8P	A194	Hot or cold forged nuts for high pressure & high temperature service	AISI 305 (with restricted carbon)	Yes	⅛ and larger	—	—	—	80,000 / 75,000	126/300	B60/B105	—
8PB	A194	Nuts machined from bars for high pressure & high temperature service	AISI 305 (with restricted carbon)	Yes	⅛ and larger	—	—	—	80,000 / 75,000	126/300	B60/B105	—
8PA	A194	Hot or cold forged or machined from bars for high pressure & high temperature service	AISI 305 (with restricted carbon), carbide solution treated	Yes	⅛ and larger	—	—	—	80,000 / 75,000	126/192	B60/B90	—
8N	A194	Hot or cold forged nuts for high pressure & high temperature service	AISI 304N	Yes	⅛ and larger	—	—	—	80,000 / 75,000	126/300	B60/B105	—
8NB	A194	Nuts machined from bars for high pressure & high temperature service	AISI 304N	Yes	⅛ and larger	—	—	—	80,000 / 75,000	126/300	B60/B105	—
8NA	A194	Hot or cold forged or machined from bars for high pressure & high temperature service	AISI 304N, carbide solution treated	Yes	⅛ and larger	—	—	—	80,000 / 75,000	126/192	B60/B90	—
8MN	A194	Hot or cold forged nuts for high pressure & high temperature service	AISI 316N	Yes	⅛ and larger	—	—	—	80,000 / 75,000	126/300	B60/B105	—
8MNB	A194	Nuts machined from bars for high pressure & high temperature service	AISI 316N	Yes	⅛ and larger	—	—	—	80,000 / 75,000	126/300	B60/B105	—
8MNA	A194	Hot or cold forged or machined from bars for high pressure & high temperature service	AISI 316N, carbide solution treated	Yes	⅛ and larger	—	—	—	80,000 / 75,000	126/192	B60/B90	—
8R	A194	Hot or cold forged nuts for high pressure & high temperature service	XM19	Yes	⅛ and larger	—	—	—	80,000 / 75,000	183/271	B88/C25	—
8RB	A194	Nuts machined from bars for high pressure & high temperature service	XM19	Yes	⅛ and larger	—	—	—	80,000 / 75,000	183/271	B88/C25	—

Footnotes are grouped on the last page of this Part 1 series.

Grade ID mark	ASTM spec number	Fastener description	Material	Is mfgr's ID symbol req'd?	Nominal size range (inch)	Mechanical properties: Bolts, screws, studs: Proof load (psi)	Bolts, screws, studs: Yield strength (min psi)	Bolts, screws, studs: Tensile strength (min psi)	Nuts: Proof load (psi) hvy hex/hex	Hardness: Brinell	Hardness: Rockwell	Remarks or footnotes
8RA	A194	Hot or cold forged or machined from bars for high pressure & high temperature service	XM19, carbide solution treated	Yes	¼ and larger	—	—	—	80,000/75,000	183/271	B88/C25	—
8S	A194	Hot or cold forged nuts for high pressure & high temperature service	S21800 (restricted phosphorus)	Yes	¼ and larger	—	—	—	80,000/75,000	183/271	B88/C25	—
8SB	A194	Nuts machined from bars for high pressure & high temperature service	S21800 (restricted phosphorus)	Yes	¼ and larger	—	—	—	80,000/75,000	183/271	B88/C25	—
8SA	A194	Hot or cold forged or machined from bars for high pressure & high temperature service	S21800 (restricted phosphorus), carbide solution treated	Yes	¼ and larger	—	—	—	80,000/75,000	183/271	B88/C25	—
8LN	A194	Hot or cold forged nuts for high pressure & high temperature service	AISI 304N (with restricted carbon)	Yes	¼ and larger	—	—	—	80,000/75,000	126/300	B60/B105	—
8LNB	A194	Nuts machined from bars for high pressure & high temperature service	AISI 304N (with restricted carbon)	Yes	¼ and larger	—	—	—	80,000/75,000	126/300	B60/B105	—
8LNA	A194	Hot or cold forged or machined from bars for high pressure & high temperature service	AISI 304N (with restricted carbon), carbide solution treated	Yes	¼ and larger	—	—	—	80,000/75,000	126/192	B60/B90	—
8MLN	A194	Hot or cold forged nuts for high pressure & high temperature service	AISI 316N (with restricted carbon)	Yes	¼ and larger	—	—	—	80,000/75,000	126/300	B60/B105	—
8MLNB	A194	Nuts machined from bars for high pressure & high temperature service	AISI 316N (with restricted carbon	Yes	¼ and larger	—	—	—	80,000/75,000	126/300	B60/B105	—
8MLNA	A194	Hot or cold forged or machined from bars for high pressure & high temperature service	AISI 316N (with restricted carbon), carbide solution treated	Yes	¼ and larger	—	—	—	80,000/75,000	126/192	B60/B90	—
8	A194	Nuts machined from bars for high pressure & high temperature service	AISI 304, strain hardened	Yes	¼ - ¾	—	—	—	125,000/110,000	—	—	—
					⅞ - 1	—	—	—	115,000/100,000	—	—	—
					1⅛ - 1¼	—	—	—	105,000/95,000	—	—	—
					1⅜ - 1½	—	—	—	100,000/90,000	—	—	—

Footnotes are grouped on the last page of this Part 1 series.

Grade ID mark	ASTM spec number	Fastener description	Material	Is mfgr's ID symbol req'd?	Nominal size range (inch)	Mechanical properties				Hardness		Remarks or footnotes
						Bolts, screws, studs			Nuts			
						Proof load (psi)	Yield strength (min psi)	Tensile strength (min psi)	Proof load (psi) hvy hex / hex	Brinell	Rockwell	
8C	A194	Nuts machined from bars for high pressure & high temperature service	AISI 347, strain hardened	Yes	1/4 - 3/4	—	—	—	125,000 / 110,000	—	—	—
					3/4 - 1	—	—	—	115,000 / 100,000	—	—	—
					1 1/8 - 1 1/4	—	—	—	105,000 / 95,000	—	—	—
					1 3/8 - 1 1/2	—	—	—	100,000 / 90,000	—	—	—
8T	A194	Nuts machined from bars for high pressure & high temperature service	AISI 321, strain hardened	Yes	1/4 - 3/4	—	—	—	125,000 / 110,000	—	—	—
					3/4 - 1	—	—	—	115,000 / 100,000	—	—	—
					1 1/8 - 1 1/4	—	—	—	105,000 / 95,000	—	—	—
					1 3/8 - 1 1/2	—	—	—	100,000 / 90,000	—	—	—
8M	A194	Nuts machined from bars for high pressure & high temperature service	AISI 316, strain hardened	Yes	1/4 - 3/4	—	—	—	110,000 / 100,000	—	—	—
					3/4 - 1	—	—	—	100,000 / 90,000	—	—	—
					1 1/8 - 1 1/4	—	—	—	95,000 / 85,000	—	—	—
					1 3/8 - 1 1/2	—	—	—	90,000 / 80,000	—	—	—
8F	A194	Nuts machined from bars for high pressure & high temperature service	AISI 303 with sulfur or 303Se with selenium, strain hardened	Yes	1/4 - 3/4	—	—	—	125,000 / 110,000	—	—	—
					3/4 - 1	—	—	—	115,000 / 100,000	—	—	—
					1 1/8 - 1 1/4	—	—	—	105,000 / 95,000	—	—	—
					1 3/8 - 1 1/2	—	—	—	100,000 / 90,000	—	—	—
8P	A194	Nuts machined from bars for high pressure & high temperature service	AISI 305 (with restricted carbon), strain hardened	Yes	1/4 - 3/4	—	—	—	125,000 / 110,000	—	—	—
					3/4 - 1	—	—	—	115,000 / 100,000	—	—	—
					1 1/8 - 1 1/4	—	—	—	105,000 / 95,000	—	—	—
					1 3/8 - 1 1/2	—	—	—	100,000 / 90,000	—	—	—

Footnotes are grouped on the last page of this Part 1 series.

Grade ID mark	ASTM spec number	Fastener description	Material	Is mfgr's ID symbol req'd?	Nominal size range (inch)	Bolts, screws, studs Proof load (psi)	Yield strength (min psi)	Tensile strength (min psi)	Nuts Proof load (psi) hvy hex	hex	Brinell	Rockwell	Remarks or footnotes
8N	A194	Nuts machined from bars for high pressure & high temperature service	AISI 304N, strain hardened	Yes	1/4 - 7/8	—	—	—	125,000	110,000	—	—	—
					7/8 - 1	—	—	—	115,000	100,000	—	—	—
					1 1/8 - 1 1/4	—	—	—	105,000	95,000	—	—	—
					1 3/8 - 1 1/2	—	—	—	100,000	90,000	—	—	—
8MN	A194	Nuts machined from bars for high pressure & high temperature service	AISI 316N, strain hardened	Yes	1/4 - 7/8	—	—	—	125,000	110,000	—	—	—
					7/8 - 1	—	—	—	115,000	100,000	—	—	—
					1 1/8 - 1 1/4	—	—	—	105,000	95,000	—	—	—
					1 3/8 - 1 1/2	—	—	—	100,000	90,000	—	—	—
None req'd	A307, Grade A	Common bolts	Carbon steel	Yes	1/4 - 4	—	—	60,000	—	—	121/241 F	B69/B100 F	Marking on head, raised or depressed.
None req'd	A307, Grade B	Bolts for flanged joints	Carbon steel	Yes	1/4 - 4	—	—	60,000 min 100,000 max	—	—	121/212	B69/B95	Marking on head, raised or depressed.
L7	A320	Bolts, screws, and studs for low temperature service	AISI 4140, 4142, or 4145 quenched & tempered	Yes	1/4 - 2 1/2	—	105,000	125,000	—	—	—	—	A
L7A	A320	Bolts, screws, and studs for low temperature service	AISI 4037 quenched & tempered	Yes	1/4 - 2 1/2	—	105,000	125,000	—	—	—	—	A
L7B	A320	Bolts, screws, and studs for low temperature service	AISI 4137 quenched & tempered	Yes	1/4 - 2 1/2	—	105,000	125,000	—	—	—	—	A
L7C	A320	Bolts, screws, and studs for low temperature service	AISI 8740 quenched & tempered	Yes	1/4 - 2 1/2	—	105,000	125,000	—	—	—	—	A
L70	A320	Bolts, screws, and studs for low temperature service	AISI 4140, 4142, or 4145 quenched & tempered	Yes	1/4 - 2 1/2	—	105,000	125,000	—	—	—	—	A
L71	A320	Bolts, screws, and studs for low temperature service	AISI 4037 quenched & tempered	Yes	1/4 - 2 1/2	—	105,000	125,000	—	—	—	—	A
L72	A320	Bolts, screws, and studs for low temperature service	AISI 4137 quenched & tempered	Yes	1/4 - 2 1/2	—	105,000	125,000	—	—	—	—	A

Footnotes are grouped on the last page of this Part 1 series.

Grade ID mark	ASTM spec number	Fastener description	Material	Is mfgr's ID symbol req'd?	Nominal size range (inch)	Bolts, screws, studs			Nuts	Hardness		Remarks or footnotes
						Proof load (psi)	Yield strength (min psi)	Tensile strength (min psi)	Proof load (psi)	Brinell	Rockwell	
L73	A320	Bolts, screws, and studs for low temperature service	AISI 8740 quenched & tempered	Yes	¼ - 2½	—	105,000	125,000	—	—	—	A
L43	A320	Bolts, screws, and studs for low temperature service	AISI 4340 quenched & tempered	Yes	¼ - 4	—	105,000	125,000	—	—	—	A
L7M	A320	Bolts, screws, and studs for low temperature service	AISI 4140, 4142, or 4145 quenched & tempered	Yes	¼ - 2½	—	80,000	100,000	—	235^G max	B99^G max	A
L1	A320	Bolts, screws, and studs for low temperature service	Low carbon martensite steel, quenched & tempered	Yes	¼ - 1	—	105,000	125,000	—	—	—	A
B8F	A320	Bolts, screws, and studs for low temperature service, Class 1	AISI 303 with sulfur or 303Se with selenium, carbide solution treated	Yes	¼ and larger	—	30,000	75,000	—	223^D max	B96^D max	A
B8FA	A320	Bolts, screws, and studs for low temperature service, Class 1A	AISI 303 or 303Se carbide solution treated in finished condition	Yes	¼ and larger	—	30,000	75,000	—	192 max	B90 max	A
B8F	A320	Bolts, screws, and studs for low temperature service, Class 2	AISI 303 or 303Se carbide solution treated and strain hardened	Yes	¼ - ¾	—	100,000	125,000	—	321 max	C35 max	A
					Over ¾ - 1	—	80,000	115,000	—	321 max	C35 max	A
					Over 1 - 1¼	—	65,000	105,000	—	321 max	C35 max	A
					Over 1¼ - 1½	—	50,000	100,000	—	321 max	C35 max	A
A325 or option ＼／ A325	A325, Type 1	High strength structural bolts	Medium carbon steel, quenched & tempered	Yes	½ - 1	85,000	92,000	120,000	—	248/331	C24/C35	H,I
					1⅛ - 1½	74,000	81,000	105,000	—	223/293	C19/C31	H,I
＼／ A325	A325, Type 2	High strength structural bolts	Low carbon martensite steel, quenched & tempered	Yes	½ - 1	85,000	92,000	120,000	—	248/331	C24/C35	H,I
					1⅛ - 1½	74,000	81,000	105,000	—	223/293	C19/C31	H,I
A325	A325, Type 3	High strength structural bolts	Weathering steel, quenched & tempered	Yes	½ - 1	85,000	92,000	120,000	—	248/331	C24/C35	H, I, J
					1⅛ - 1½	74,000	81,000	105,000	—	223/293	C19/C31	H, I, J
A325M 8S	A325M, Type 1	High strength structural bolts—metric	Medium carbon steel, quenched & tempered	Yes	M16 - M36	600 MPa	660 MPa	830 MPa	—	Vickers 255/336	C23/C34	K, L
A325M 8S	A325M, Type 2	High strength structural bolts—metric	Low carbon martensite steel, quenched & tempered	Yes	M16 - M36	600 MPa	660 MPa	830 MPa	—	Vickers 255/336	C23/C34	K, L

Mechanical properties

Footnotes are grouped on the last page of this Part 1 series.

48

Grade ID mark	ASTM spec number	Fastener description	Material	Is mfgr's ID symbol req'd?	Nominal size range (inch)	Mechanical properties: Bolts, screws, studs: Proof load (psi)	Bolts, screws, studs: Yield strength (min psi)	Bolts, screws, studs: Tensile strength (min psi)	Nuts: Proof load (psi)	Hardness: Brinell	Hardness: Rockwell	Remarks or footnotes
A325M 8S3	A325M, Type 3	High strength structural bolts—metric	Weathering steel, quenched & tempered	Yes	M16 - M36	600 MPa	660 MPa	830 MPa	—	Vickers 255/336	C23/C34	J, K, L
BC	A354, Grade BC	Bolts & studs	Alloy steel, quenched & tempered	Yes	¼ - 2½ Over 2½ - 4	105,000 95,000	109,000 99,000	125,000 115,000	— —	255/331 235/311	C26/36 C22/C33	H,N H,N
BD or  (M)	A354, Grade BD	Bolts & studs	Alloy steel, quenched & tempered	Yes	¼ - 2½ Over 2½ - 4	120,000 105,000	130,000 115,000	150,000 140,000	— —	311/363 293/363	C33/C39 C31/C39	H,N,O H,N,O
None req'd (P)	A394	Transmission tower bolts	Galvanized steel	Yes	½, ⅝, ¾, ⅞, 1	(Single shear at threads based on 45,000 psi.)		60,000	—	121/235	B69/B99	Marking on head, raised or depressed. (H)
B4B	A437, Grade B4B	Turbine-type bolts, screws, studs, nuts, and washers for high temperature service	Alloy steel, specially heat treated	Yes	All dia	—	105,000	145,000	—	See remarks	C31/C37 for nuts & washers	331 max for bolts & studs; 293/341 for nuts & washers. (A)
B4C	A437, Grade B4C	Turbine-type bolts, screws, studs, nuts, and washers for high temperature service	Alloy steel, specially heat treated	Yes	All dia	—	85,000	115,000	—	See remarks	C21/C29 for nuts & washers	277 max for bolts & studs; 229/277 for nuts & washers. (A)
B4D	A437, Grade B4D	Turbine-type bolts, screws, studs, nuts, and washers for high temperature service	Alloy steel, specially heat treated	Yes	Thru 2½ Over 2½ - 4 Over 4 - 7	— — —	105,000 95,000 85,000	125,000 110,000 100,000	— — —	See remarks See remarks	C27/C33 for nuts and washers	302 max for bolts and studs; 263/311 for nuts and washers. (A)
	A449	Bolts and studs	Medium carbon steel, quenched & tempered	Yes	¼ - 1 Over 1 - 1½ Over 1½ - 3	85,000 74,000 55,000	92,000 81,000 58,000	120,000 105,000 90,000	— — —	255/321 223/285 183/235	C25/C34 C19/C30 —	Marking on head, raised or depressed. (H,Q)
660A (R)	A453, Grade 660 Class A	Bolts, screws, studs, nuts, and washers for high temperature service	Special alloy steel, specially heat treated	Yes	¼ and larger	—	85,000	130,000	—	248/341	—	A,R
660B (R)	A453, Grade 660 Class B	Bolts, screws, studs, nuts, and washers for high temperature service	Special alloy steel, specially heat treated	Yes	¼ and larger	—	85,000	130,000	—	248/341	—	A,R

Footnotes are grouped on the last page of this Part 1 series.

Grade ID mark	ASTM spec number	Fastener description	Material	Is mfgr's ID symbol req'd?	Nominal size range (inch)	Mechanical properties: Bolts, screws, studs: Proof load (psi)	Bolts, screws, studs: Yield strength (min psi)	Bolts, screws, studs: Tensile strength (min psi)	Nuts: Proof load (psi)	Hardness: Brinell	Hardness: Rockwell	Remarks or footnotes
651A (R)	A453, Grade 651 Class A	Bolts, screws, studs, nuts, and washers for high temperature service	Special alloy steel, specially heat treated	Yes	¼ - 3 Over 3 and larger	— —	70,000 60,000	100,000 100,000	— —	220/280 220/280	— —	A,R A,R
651B (R)	A453, Grade 651 Class B	Bolts, screws, studs, nuts, and washers for high temperature service	Special alloy steel, specially heat treated	Yes	¼ - 3 Over 3 and larger	— —	60,000 50,000	95,000 95,000	— —	210/270 210/270	— —	A,R A,R
662A (R)	A453, Grade 662 Class A	Bolts, screws, studs, nuts, and washers for high temperature service	Special alloy steel, specially heat treated	Yes	¼ and larger	—	85,000	130,000	—	255/321	—	A,R
662B (R)	A453, Grade 662 Class B	Bolts, screws, studs, nuts, and washers for high temperature service	Special alloy steel, specially heat treated	Yes	¼ and larger	—	80,000	125,000	—	248/321	—	A
665A (R)	A453, Grade 665 Class A	Bolts, screws, studs, nuts, and washers for high temperature service	Special alloy steel, specially heat treated	Yes	¼ and larger	—	120,000	170,000	—	311/388	—	A
665B (R)	A453, Grade 665 Class B	Bolts, screws, studs, nuts, and washers for high temperature service	Special alloy steel, specially heat treated	Yes	¼ and larger	—	120,000	155,000	—	311/388	—	A
None req'd	A489	Eyebolts	Carbon steel, quenched and tempered	Yes	¼ - 2½	(S)	30,000	65,000 min, 85,000 max	—	—	—	T

ASTM footnotes

A. Grade and manufacturer's identification symbols shall be applied to one end of studs ¼" in diameter and larger, and to the heads of bolts and screws ¼" in diameter and larger. (If available area is inadequate, grade symbol may be marked on one end and manufacturer's identification symbol marked on the other end.)

B. To meet tensile requirements, Brinell hardness shall be over 201 HB (94 HRB)

C. A193 products are for high temperature service, A320 products are for low temperature service.

D. For sizes ¼" in diameter and smaller, maximum hardness of 241 HB (100 HRB) is permitted

E. Nuts machined from heat treated bars need not be re-heat treated

F. Except when tested by wedge tension test.

G. To meet tensile requirements, Brinell hardness shall not be less than 200 HB or 93 HRB

H. Bolts (screws) less than three diameters in length (and studs less than four diameters in length) shall have hardness values not less than the minimum nor more than the maximum hardness limits required, as hardness is their only mechanical requirement

I. Excluding studs, all markings located on top of head, raised or depressed.

J. Manufacturer may add other distinguishing marks indicating the fastener is atmospheric corrosion resistant and of a weathering type

K. All markings shall be located on top of the [illegible] raised or depressed. Base of property class symbol [illegible] be positioned toward the closest periphery of the [illegible]

L. Short length bolts need only meet hardness limits, as hardness is their only mechanical requirement. (Refer to Table 4 of ASTM F568 for definition of minimum length of product subject to tensile testing.)

M. Grade BD bolts ¼" through 1½" diameter shall be marked with six radial lines 60 degrees apart on top of bolt head, instead of the grade symbol.

N. Marks may be raised or depressed on the top of the head for bolts and on one end for studs.

O. Grade BD in sizes ¼" through 1½" is equivalent to SAE Grade 8. (Note: AISI 1541 does not satisfy chemical requirements for Grade BD.)

P. Bolts may be marked on the head by agreement with some identifying symbol to indicate special material and processing requirements when applicable

Q. A449 in sizes ¼" through 1½" is equivalent to SAE Grade 5

R. In addition to identification symbols (grade and class), the type designation 2 shall also appear on all roll-threaded bolting material so processed. Absence of the type designation number indicates Type 1 processed material or machine-cut threads

S. Refer to ASTM Standard A489 for specific strength requirements

T. Manufacturer's name or identification mark shall be forged in raised characters on eyebolt surface

Grade and material markings—Part II

ASTM markings

The American Society for Testing and Materials, 1916 Race St Philadelphia PA 19103, sponsors development of specifications for fasteners used in general and special engineering applications. These specifications detail chemical and mechanical properties of material strength levels for fasteners and are generally specific in referencing the actual product covered. A full range of types of products of various styles, thread series, lengths, etc. can be produced to meet ASTM requirements and would be marked for grade and material identification as required.

ASTM Grade and material identification markings required by ASTM specifications

Grade ID mark	ASTM spec number	Fastener description	Material	Is mfgr's ID symbol req'd?	Nominal size range (inch)	Bolts, screws, studs			Nuts	Hardness		Remarks or footnotes
						Proof load (psi)	Yield strength (min psi)	Tensile strength (min psi)	Proof load (psi)	Brinell	Rockwell	
A490	A490, Type 1	High strength structural bolts	Alloy steel, quenched & tempered	Yes	½ - 1½	120,000	130,000	150,000 min 170,000 max	—	311/352	C33/C38	Marking on top of head, raised or depressed. (H)
⚡ A490	A490, Type 2	High strength structural bolts	Low carbon martensite steel, quenched & tempered	Yes	½ - 1½	120,000	130,000	150,000 min 170,000 max	—	311/352	C33/C38	Marking on top of head, raised or depressed. (H)
A490	A490, Type 3	High strength structural bolts	Weathering steel, quenched & tempered	Yes	½ - 1½	120,000	130,000	150,000 min 170,000 max	—	311/352	C33/C38	Marking on top of head, raised or depressed. (H,J)
A490M 10S	A490M, Type 1	High strength structural bolts—metric	Alloy steel, quenched & tempered	Yes	M16 - M36 mm	830 MPa	940 MPa	1040 MPa	—	Vickers 327/382	C33/C39	K
A490M 10S	A490M, Type 2	High strength structural bolts—metric	Low carbon martensite steel, quenched & tempered	Yes	M16 - M36 mm	830 MPa	940 MPa	1040 MPa	—	Vickers 327/382	C33/C39	K
A490M 10S3	A490M, Type 3	High strength structural bolts—metric	Weathering steel, quenched & tempered	Yes	M16 - M36 mm	830 MPa	940 MPa	1040 MPa	—	Vickers 327/382	C33/C39	J,K
None req'd (U)	A502, Grade 1	Structural rivets	Carbon steel	Yes	½ - 1½	—	—	—	—	103/126	B55/B72	Markings on top of rivet head, raised or depressed
2	A502, Grade 2	Structural rivets	Carbon manganese steel	Yes	½ - 1½	—	—	—	—	137/163	B76/B85	Markings on top of rivet head, raised or depressed
3	A502, Grade 3	Structural rivets	Weathering steel	Yes	½ - 1½	—	—	—	—	137/197	B76/B93	Markings on top of rivet head, raised or depressed

Footnotes are grouped on the last page of this Part II series.

Grade ID mark	ASTM spec number	Fastener description	Material	Is mfgr's ID symbol req'd?	Nominal size range (inch)	Bolts, screws, studs			Nuts	Hardness		Remarks or footnotes
						Proof load (psi)	Yield strength (min psi)	Tensile strength (min psi)	Proof load (psi)	Brinell	Rockwell	
B21 (V)	A540, Grade B21, Class 5	Bolts, studs, washers, and nuts for nuclear and other special applications	Alloy steel (Cr-Mo-V), quenched & tempered	Yes	Thru 2	—	105,000	120,000	—	241/285	—	W
					Over 2 - 6	—	100,000	115,000	—	248/302	—	W
					Over 6 - 8	—	100,000	115,000	—	255/311	—	W
B21 (V)	A540, Grade B21, Class 4	Bolts, studs, washers, and nuts for nuclear and other special applications	Alloy steel (Cr-Mo-V), quenched & tempered	Yes	Thru 3	—	120,000	135,000	—	269/331	—	W
					Over 3 - 6	—	120,000	135,000	—	277/352	—	W
B21 (V)	A540, Grade B21, Class 3	Bolts, studs, washers, and nuts for nuclear and other special applications	Alloy steel (Cr-Mo-V), quenched & tempered	Yes	Thru 3	—	130,000	145,000	—	293/352	—	W
					Over 3 - 6	—	130,000	145,000	—	302/375	—	W
B21 (V)	A540, Grade B21, Class 2	Bolts, studs, washers, and nuts for nuclear and other special applications	Alloy steel (Cr-Mo-V), quenched & tempered	Yes	Thru 4	—	140,000	155,000	—	311/401	—	W
B21 (V)	A540, Grade B21, Class 1	Bolts, studs, washers, and nuts for nuclear and other special applications	Alloy steel (Cr-Mo-V), quenched & tempered	Yes	Thru 4	—	150,000	165,000	—	321/429	—	W
B22 (V)	A540, Grade B22, Class 5	Bolts, studs, washers, and nuts for nuclear and other special applications	AISI 4142-H, quenched & tempered	Yes	Thru 2	—	105,000	120,000	—	248/293	—	W
					Over 2 - 4	—	100,000	115,000	—	255/302	—	W
B22 (V)	A540, Grade B22, Class 4	Bolts, studs, washers, and nuts for nuclear and other special applications	AISI 4142-H, quenched & tempered	Yes	Thru 1	—	120,000	135,000	—	269/341	—	W
					Over 1 - 4	—	120,000	135,000	—	277/363	—	W
B22 (V)	A540, Grade B22, Class 3	Bolts, studs, washers, and nuts for nuclear and other special applications	AISI 4142-H, quenched & tempered	Yes	Thru 2	—	130,000	145,000	—	293/363	—	W
					Over 2 - 4	—	130,000	145,000	—	302/375	—	W
B22 (V)	A540, Grade B22, Class 2	Bolts, studs, washers, and nuts for nuclear and other special applications	AISI 4142-H, quenched & tempered	Yes	Thru 3	—	140,000	155,000	—	311/401	—	W
B22 (V)	A540, Grade B22, Class 1	Bolts, studs, washers, and nuts for nuclear and other special applications	AISI 4142-H, quenched & tempered	Yes	Thru 1½	—	150,000	165,000	—	321/401	—	W
B23 (V)	A540, Grade B23, Class 5	Bolts, studs, washers, and nuts for nuclear and other special applications	AISI E-4340-H, quenched & tempered	Yes	Thru 6	—	105,000	120,000	—	248/311	—	W
					Over 6 - 8	—	100,000	115,000	—	255/321	—	W
					Over 8 - 9½	—	100,000	115,000	—	262/321	—	W

Footnotes are grouped on the last page of this Part II series

Grade ID mark	ASTM spec number	Fastener description	Material	Is mfgr's ID symbol req'd?	Nominal size range (inch)	Bolts, screws, studs			Nuts	Hardness		Remarks or footnotes
						Proof load (psi)	Yield strength (min psi)	Tensile strength (min psi)	Proof load (psi)	Brinell	Rockwell	
B23 (V)	A540, Grade B23, Class 4	Bolts, studs, washers, and nuts for nuclear and other special applications	AISI E-4340-H, quenched & tempered	Yes	Thru 3	—	120,000	135,000	—	269/341	—	W
					Over 3 - 6	—	120,000	135,000	—	277/352	—	W
					Over 6 - 9½	—	120,000	135,000	—	285/363	—	W
B23 (V)	A540, Grade B23, Class 3	Bolts, studs, washers, and nuts for nuclear and other special applications	AISI E-4340-H, quenched & tempered	Yes	Thru 3	—	130,000	145,000	—	293/363	—	W
					Over 3 - 6	—	130,000	145,000	—	302/375	—	W
					Over 6 - 9½	—	130,000	145,000	—	311/388	—	W
B23 (V)	A540, Grade B23, Class 2	Bolts, studs, washers, and nuts for nuclear and other special applications	AISI E-4340-H, quenched & tempered	Yes	Thru 3	—	140,000	155,000	—	311/388	—	W
					Over 3 - 6	—	140,000	155,000	—	311/401	—	W
					Over 6 - 9½	—	140,000	155,000	—	321/415	—	W
B23 (V)	A540, Grade B23, Class 1	Bolts, studs, washers, and nuts for nuclear and other special applications	AISI E-4340-H, quenched & tempered	Yes	Thru 3	—	150,000	165,000	—	321/415	—	W
					Over 3 - 6	—	150,000	165,000	—	331/429	—	W
					Over 6 - 8	—	150,000	165,000	—	341/444	—	W
B24 (V)	A540, Grade B24, Class 5	Bolts, studs, washers, and nuts for nuclear and other special applications	AISI 4340 Mod. quenched & tempered	Yes	Thru 6	—	105,000	120,000	—	248/311	—	W
					Over 6 - 8	—	100,000	115,000	—	255/321	—	W
					Over 8 - 9½	—	100,000	115,000	—	262/321	—	W
B24 (V)	A540, Grade B24, Class 4	Bolts, studs, washers, and nuts for nuclear and other special applications	AISI 4340 Mod. quenched & tempered	Yes	Thru 3	—	120,000	135,000	—	269/341	—	W
					Over 3 - 6	—	120,000	135,000	—	277/352	—	W
					Over 6 - 8	—	120,000	135,000	—	285/363	—	W
					Over 8 - 9½	—	120,000	135,000	—	293/363	—	W
B24 (V)	A540, Grade B24, Class 3	Bolts, studs, washers, and nuts for nuclear and other special applications	AISI 4340 Mod. quenched & tempered	Yes	Thru 3	—	130,000	145,000	—	293/363	—	W
					Over 3 - 8	—	130,000	145,000	—	302/388	—	W
					Over 8 - 9½	—	130,000	145,000	—	311/388	—	W
B24 (V)	A540, Grade B24, Class 2	Bolts, studs, washers, and nuts for nuclear and other special applications	AISI 4340 Mod. quenched & tempered	Yes	Thru 7	—	140,000	155,000	—	311/401	—	W
					Over 7 - 9½	—	140,000	155,000	—	321/415	—	W
B24 (V)	A540, Grade B24, Class 1	Bolts, studs, washers, and nuts for nuclear and other special applications	AISI 4340 Mod. quenched & tempered	Yes	Thru 6	—	150,000	165,000	—	321/415	—	W
					Over 6	—	150,000	165,000	—	331/429	—	W

Footnotes are grouped on the last page of this Part II series.

Grade ID mark	ASTM spec number	Fastener description	Material	Is mfgr's ID symbol req'd?	Nominal size range (inch)	Bolts, screws, studs Proof load (psi)	Yield strength (min psi)	Tensile strength (min psi)	Nuts Proof load (psi)	Brinell	Rockwell	Remarks or footnotes
B24V (V)	A540, Grade B24V Class 3	Bolts, studs, washers, and nuts for nuclear and other special applications	AISI 4340V Mod. quenched & tempered	Yes	Thru 4	—	130,000	145,000	—	293/363	—	W
					Over 4 - 8	—	130,000	145,000	—	302/375	-.	W
					Over 8 - 11	—	130,000	145,000	—	311/388	—	W
B24V (V)	A540, Grade B24V Class 2	Bolts, studs, washers, and nuts for nuclear and other special applications	AISI 4340V Mod. quenched & tempered	Yes	Thru 4	—	140,000	155,000	—	311/388	—	W
					Over 4 - 8	—	140,000	155,000	—	311/401	—	W
					Over 8 - 11	—	140,000	155,000	—	321/415	—	W
B24V (V)	A540, Grade B24V Class 1	Bolts, studs, washers, and nuts for nuclear and other special applications	AISI 4340V Mod. quenched & tempered	Yes	Thru 4	—	150,000	165,000	—	321/415	—	W
					Over 4 - 8	—	150,000	165,000	—	331/429	—	W
					Over 8 - 11	—	150,000	165,000	—	331/444	—	W
None req'd (X)	A563, Grade O	Nuts for general structural and mechanical use	Carbon steel	No	½ - 1½	—	—	—	Y	103/302	B55/C32	—
None req'd (X)	A563, Grade A	Nuts for general structural and mechanical use	Carbon steel	No	½ - 4	—	—	—	Y	116/302	B68/C32*	—
None req'd (X)	A563, Grade B	Nuts for general structural and mechanical use	Carbon steel	No	½ - 1	—	—	—	Y	121/302	B69/C32	—
					1½ - 1½	—	—	—	Y	121/302	B69/C32	—
( )	A563, Grade C	Heavy hex nuts for general structural and mechanical use	Carbon steel	Yes	½ - 4	—	—	—	144,000	143/352	B78/C38	Grade mark shall be applied to one nut face. (Z,AA)
(3)	A563, Grade C3	Heavy hex nuts for general structural and mechanical use	Weathering steel	Yes	½ - 4	—	—	—	144,000	143/352	B78/C38	Grade mark shall be applied to one nut face. (Z,J)
D	A563, Grade D	Nuts for general structural and mechanical use	Alloy steel	Yes	½ - 4	—	—	—	Y	159/352	B84/C38	Grade mark shall be applied to one nut face. (Z,AA)
DH	A563, Grade DH	Nuts for general structural and mechanical use	Alloy steel, quenched & tempered	Yes	½ - 4	—	—	—	Y	248/352	C24/C38	Grade mark shall be applied to one nut face. (Z,BB)
DH3	A563, Grade DH3	Heavy hex nuts for general structural and mechanical use	Weathering steel, quenched & tempered	Yes	½ - 4	—	—	—	175,000	248/352	C24/C38	Grade mark shall be applied to one nut face. (J,Z)

Footnotes are grouped on the last page of this Part II series.

54

Grade ID mark	ASTM spec number	Fastener description	Material	Is mfgr's ID symbol req'd?	Nominal size range (mm)	Mechanical properties						Remarks or footnotes
						Bolts, screws, studs			Nuts	Hardness		
						Proof load	Yield strength	Tensile strength	Proof load	Brinell	Rockwell	
5	A563M, Class 5	Nuts for general structural and mechanical use—metric	Carbon steel	No	M1.6 - M4	—	—	—	520 MPa	Vickers 130/302	B70/C30	Z,DD
					M5 & M6	—	—	—	580[CC] MPa	Vickers 130/302	B70/C30	Z,DD
					M8 & M10	—	—	—	590[CC] MPa	Vickers 130/302	B70/C30	Z,DD
					M12 - M16	—	—	—	610[CC] MPa	Vickers 130/302	B70/C30	Z,DD
					M20 - M36	—	—	—	630[CC] MPa	Vickers 130/302	B78/C30	Z,DD
					M42 - M100	—	—	—	630[CC] MPa	Vickers 146/302	B70/C30	Z,DD
						—	—	—	MPa	Vickers 128/302		
9	A563M, Class 9	Nuts for general structural and mechanical use—metric	Carbon steel	No	M3 - M4	—	—	—	900 MPa	Vickers 170/302	B85/C30	Z,DD
					M5 & M6	—	—	—	915 MPa	Vickers 188/302	B89/C30	Z,DD
					M8 & M10	—	—	—	940 MPa	Vickers 188/302	B89/C30	Z,DD
					M12 - M16	—	—	—	950 MPa	Vickers 188/302	B89/C30	Z,DD
					M20 - M100	—	—	—	920 MPa	Vickers 188/302	B89/C30	Z,DD
10	A563M, Class 10	Nuts for general structural and mechanical use—metric	Alloy steel, quenched & tempered	Yes	M1.6 - M10	—	—	—	1040 MPa	Vickers 272/353	C26/C36	Z,DD
					M12 - M16	—	—	—	1050 MPa	Vickers 272/353	C26/C36	Z,DD
					M20 - M36	—	—	—	1060 MPa	Vickers 272/353	C26/C36	Z,DD
12	A563M Class 12	Nuts for general structural and mechanical use—metric	Alloy steel, quenched & tempered	Yes	M3 - M6	—	—	—	1150[CC] MPa	Vickers 272/353	C26/C36	Z,DD
					M8 & M10	—	—	—	1160[CC] MPa	Vickers 272/353	C26/C36	Z,DD
					M12 - M16	—	—	—	1190[CC] MPa	Vickers 272/353	C26/C36	Z,DD
					M20 - M100	—	—	—	1200[CC] MPa	Vickers 272/353	C26/C36	Z,DD
8S	A563M, Class 8S	Nuts for general structural and mechanical use—metric	Carbon steel	Yes	M12 - M36	—	—	—	1075 MPa	Vickers 188/372	B89/C38	Z,DD
8S3	A563M, Class 8S3	Nuts for general structural and mechanical use—metric	Weathering steel	Yes	M12 - M36	—	—	—	1075 MPa	Vickers 188/372	B89/C38	J,Z,DD
10S	A563M, Class 10S	Nuts for general structural and mechanical use—metric	Alloy steel, quenched & tempered	Yes	M12 - M36	—	—	—	1245[CC] MPa	Vickers 272/372	C26/C38	Z,DD
10S3	A563M, Class 10S3	Nuts for general structural and mechanical use—metric	Weathering steel, quenched & tempered	Yes	M12 - M36	—	—	—	1245 MPa	Vickers 272/372	C26/C38	J,Z,DD

Footnotes are grouped on the last page of this Part II series.

Grade ID mark	ASTM spec number	Fastener description	Material	Is mfgr's ID symbol req'd?	Nominal size range (inch)	Mechanical properties						Remarks or footnotes
						Bolts, screws, studs			Nuts	Hardness		
						Proof load (psi)	Yield strength (min psi)	Tensile strength (min psi)	Proof load (psi)	Brinell	Rockwell	
None req'd See EE	A574	Socket head cap screws	Alloy steel, quenched & tempered	No (EE)	½ and smaller ⅝-4	140,000 135,000	— 153,000 FF	180,000 170,000	— —	— —	C39/C45 C37/C45	H
12.9	A574M	Socket head cap screws, metric	Alloy steel, quenched & tempered	Yes	M1.6 - M48 mm	970 MPa	1100 MPa	1220 MPa	--	Vickers 372/434 DPH	C38/C44	GG
A687	A687	Bolts & studs	Alloy steel, quenched & tempered	No	⅜ - 3	—	105,000	150,000 max	—	—	—	Marking appears on the end of product. HH
See "Remarks"	A761	Fasteners for pipe, pipe anchors and arches	Galvanized steel	Yes	⅜ only	← See Footnote II →						↑
None req'd	C646	Drill screws for gypsum board on light-gage steel shanks	Grade 1013 to 1022 carbon steel wire in accordance with ASTM A548.	No	—	—	—	—	—	—	C45 min case hardness	—
None req'd	C893	Type G screws for gypsum board to gypsum board	Grade 1013 to 1022 carbon steel wire in accordance with ASTM A548	No	—	—	—	—	—	—	C45 min case hardness	—
None req'd	C894	Type W screws for gypsum board to wood framing	Grade 1013 to 1022 carbon steel wire in accordance with ASTM A548	No	—	—	—	—	—	—	C45 min case hardness	—
GR30	F432, Grade 30	Roof & rock bolts and accessories	Carbon steel	Yes	½ - 1	—	30,000	60,000	← See ASTM F432 →	← See ASTM F432 →		JJ
GR55	F432, Grade 55	Roof & rock bolts and accessories	Carbon steel	Yes	⅜ -1½	—	55,000	85,000	← See ASTM F432 →	← See ASTM F432 →		JJ
GR75	F432, Grade 75	Roof & rock bolts and accessories	Carbon steel	Yes	½ - 1	—	75,000	100,000	← See ASTM F432 →	← See ASTM F432 →		JJ
GR40	F432, Grade 40	Roof & rock bolts — headed deformed bars	Carbon steel	Yes	All sizes	—	← See ASTM A615 →					JJ
GR60	F432, Grade 60	Roof & rock bolts — headed deformed bars	Carbon steel	Yes	All sizes	—	← See ASTM A615 →					JJ

Footnotes are grouped on the last page of this Part II series

Grade ID mark	ASTM spec number	Fastener description	Material	Is mfgr's ID symbol req'd?	Nominal size range (inch)	Bolts, screws, studs			Nuts	Hardness		Remarks or footnotes
						Proof load (psi)	Yield strength (min psi)	Tensile strength (min psi)	Proof load (psi)	Brinell	Rockwell	
See "Remarks"	F436	Hardened washers	Carbon or weathering steel	Yes	½ - 4	—	—	—	—	—	See ASTM F436	Type 3 (weathering steel) washers shall be marked with the symbol "3". KK
M	F436M	Hardened washers—Metric Type 1	Carbon steel	Yes	M12 - M100mm	—	—	—	—	—	TT	KK
3M	F436M	Hardened washers—Metric Type 3	Weathering steel	Yes	M12 - M100mm	—	—	—	—	—	TT	KK
None req'd	F467	Nuts for general use	ETP copper UNS C11000	No	¼ - 1½	—	—	—	30,000 min	—	F65 min	—
None req'd	F467	Nuts for general use	Brass UNS 27000	No	¼ - 1½	—	—	—	60,000 min	—	F55 min	—
None req'd	F467	Nuts for general use	Naval brass UNS C46200	No	¼ - 1½	—	—	—	50,000 min	—	B65 min	—
None req'd	F467	Nuts for general use	Naval brass UNS C46400	No	¼ - 1½	—	—	—	50,000 min	—	B55 min	—
None req'd	F467	Nuts for general use	Phosphor bronze UNS C51000	No	¼ - 1½	—	—	—	60,000 min	—	B60 min	—
None req'd	F467	Nuts for general use	Aluminum bronze UNS C61300	No	¼ - 1½	—	—	—	80,000 min	—	B70 min	—
None req'd	F467	Nuts for general use	Aluminum bronze UNS C61400	No	¼ - 1½	—	—	—	75,000 min	—	B70 min	—
None req'd	F467	Nuts for general use	Aluminum bronze UNS C63000	No	¼ - 1½	—	—	—	100,000 min	—	B85 min	—
None req'd	F467	Nuts for general use	Aluminum silicon bronze UNS C64200	No	¼ - 1½	—	—	—	75,000 min	—	B75 min	—
None req'd	F467	Nuts for general use	Silicon bronze UNS C65100	No	¼ - 1½	—	—	—	70,000 min	—	B75 min	—
None req'd	F467	Nuts for general use	Silicon bronze UNS C65500	No	¼ - 1½	—	—	—	70,000 min	—	B60 min	—
None req'd	F467	Nuts for general use	Silicon bronze UNS C66100	No	¼ - 1½	—	—	—	70,000 min	—	B75 min	—
None req'd	F467	Nuts for general use	Manganese bronze UNS C67500	No	¼ - 1½	—	—	—	55,000 min	—	B60 min	—

Footnotes are grouped on the last page of this Part II series.

Grade ID mark	ASTM spec number	Fastener description	Material	Is mfgr's ID symbol req'd?	Nominal size range (inch)	Mechanical properties						Remarks or footnotes
						Bolts, screws, studs			Nuts	Hardness		
						Proof load (psi)	Yield strength (min psi)	Tensile strength (min psi)	Proof load (psi)	Brinell	Rockwell	
None req'd	F467	Nuts for general use	Cupro-nickel UNS C71000	No	¼ - 1½	—	—	—	45,000 min	—	B50 min	—
None req'd	F467	Nuts for general use	Cupro nickel UNS C71500	No	¼ - 1½	—	—	—	55,000 min	—	B60 min	—
None req'd	F467	Nuts for general use	Ni-Mo based UNS N10001	No	¼ - 1½	—	—	—	115,000 min	—	C20 min	—
None req'd	F467	Nuts for general use	Ni-Mo-Cr based UNS N10002	No	¼ - 1½	—	—	—	110,000 min	—	C20 min	—
None req'd	F467	Nuts for general use	Ni-Cu Class A UNS N04400	No	¼ - 1½	—	—	—	80,000 min	—	B75 min	—
None req'd	F467	Nuts for general use	Ni-Cu Class B UNS N04405	No	¼ - 1½	—	—	—	70,000 min	—	B60 min	—
None req'd	F467	Nuts for general use	Ni-Cu-Al based UNS N05500	No	¼ - 1½	—	—	—	130,000 min	—	C24 min	—
None req'd	F467	Nuts for general use	Aluminum 2024 UNS A92024	No	¼ - 1½	—	—	—	55,000 min	—	B70 min	—
None req'd	F467	Nuts for general use	Aluminum 6061 UNS A96061	No	¼ - 1½	—	—	—	40,000 min	—	B40 min	—
None req'd	F467	Nuts for general use	Aluminum 6262 UNS A96262	No	¼ - 1½	—	—	—	52,000 min	—	B60 min	—

ASTM footnotes

H. Bolts (screws) less than three diameters in length (and studs less than four diameters in length) shall have hardness values not less than the minimum nor more than the maximum hardness limits required, as hardness is their only mechanical requirement.

I. Excluding studs, all markings located on top of head, raised or depressed.

J. Manufacturer may add other distinguishing marks indicating the fastener is atmospheric corrosion resistant and of a weathering type.

K. All markings shall be located on top of the head, raised or depressed. Base of property class symbols shall be positioned toward the closest periphery of the head.

U. The numeral 1 may be used at manufacturer's option.

V. Marking of class identification is not mandatory.

W. Grade and manufacturer's identification symbols shall be applied to one end of studs and to heads of bolts of all sizes. (If available area is inadequate, grade symbol may be marked on one end and the manufacturer's identification symbol marked on the other end.) For bolts and studs smaller than ¼" diameter and for ¼" studs requiring more than a total of three symbols, the marking shall be a matter of agreement between purchaser and manufacturer.

X. When individual grade marking is specified in the inquiry and order, the mark shall be the grade letter symbol on one nut face.

Y. Proof load stresses vary depending on nut style, size, thread series, and in some cases on whether or not nuts are galvanized. Refer to ASTM Standard A563 for specific values.

Z. Marks may be raised or depressed. If, however, marks are located on the bearing surface (or on one of the wrenching flats for A563M faasteners) they shall be depressed.

AA. Nuts made in accordance with ASTM A194, Grade 2 or 2H, and marked with their grade symbol are acceptable equivalents for Grades C and D nuts.

BB. Nuts made in accordance with ASTM A194, Grade 2H, and marked with its grade symbol are an acceptable equivalent for Grade DH nuts.

CC. Proof load stresses are reduced for [illegible] end nuts. Refer to ASTM A563M for specific values.

DD. Nuts in nominal thread diameters M4 and smaller need not be marked. Property class designations shall be located on the top or bearing surface, on the top of flange, or on one of the wrenching flats of the nut. Markings located on the top or bearing surface or on the top of the flange shall be positioned with the base of the numeral(s) oriented toward the nut periphery. (Class 9 nuts marked on one wrenching flat shall have the numeral 9 underlined.)

EE. In the USA there is only one grade of socket head cap screw commercially available and most manufacturers apply their own source marks by knurling pattern around the outside of the head.

FF. When equipment of sufficient capacity is not readily available, machined specimens shall meet 153 ksi, min yield strength.

GG. All screws with nominal diameters of 5 mm and larger require marking. Marking may be on side or top of head.

HH. Marking small sizes (customarily less than ¼") may not be practical. Consult producer for minimum size that can be marked.

II. Assembly bolts conform to ASTM A449 requirements, nuts conform to Grade C of A563. Headwall anchorage bolting material conforms to A307 and nuts conform to Grade B of A563.

JJ. Bolt heads shall be marked with either raised or depressed marks at manufacturer's option. Marks shall include grade and diameter identification when applicable, manufacturer's symbol and bolt length.

Threaded bars, threaded slotted bars, extensions, beveled washers, threaded tapered plugs, wedges, spherical washers, threaded couplings, and flat bearing plates are not required to be marked. Deformed bearing and header plates plus hardened washers shall be marked with manufacturer's symbol. Nuts shall be marked in accordance with the ASTM specification to which they were manufactured. Expansion shells shall be marked with manufacturer's symbol and hole size for which they are intended.

KK. All marking symbols shall be depressed on one washer face.

TT. Rockwell hardness is C38-C45 through-hardened tempered. C26-C45 hot-dip galvanize. A09 A73 carburized OST

Grade and material markings —Part III

ASTM markings

The American Society for Testing and Materials, 1916 Race St. Philadelphia PA 19103, sponsors development of specifications for fasteners used in general and special engineering applications These specifications detail chemical and mechanical properties of material strength levels for fasteners and are generally specific in referencing the actual product covered. A full range of types of products of various styles, thread series, lengths, etc, can be produced to meet ASTM requirements and would be marked for grade and material identification as required.

ASTM Grade and material identification markings required by ASTM specifications

Grade ID mark	ASTM spec number	Fastener description	Material	Is mfgr's ID symbol req'd?	Nominal size range (inch)	Mechanical properties							Remarks or footnotes
						Bolts, screws, studs			Nuts	Hardness			
						Proof load (psi)	Yield strength (min psi)	Tensile strength (min psi)	Proof load (psi)	Brinell	Rockwell		
None req'd	F467	Nuts for general use	Titanium Gr 1	No	¼ - 1½	—	—	—	40,000 min	—	Vickers 140 min		—
None req'd	F467	Nuts for general use	Titanium Gr 2	No	¼ - 1½	—	—	—	55,000 min	—	Vickers 150 min		—
None req'd	F467	Nuts for general use	Titanium Gr 4	No	¼ - 1½	—	—	—	85,000 min	—	Vickers 200 min		—
None req'd	F467	Nuts for general use	Titanium Gr 5	No	¼ - 1½	—	—	—	135,000 min	—	C30 min		—
None req'd	F467	Nuts for general use	Titanium Gr 7	No	¼ - 1½	—	—	—	55,000 min	—	Vickers 160 min		—
None req'd	F467M	Nuts for general use— metric	ETP copper UNS C11000	No	M6 - M36 mm	—	—	—	205 MPa	—	F65 min		—
None req'd	F467M	Nuts for general use— metric	Brass UNS C27000	No	M6 - M36 mm	—	—	—	415 MPa	—	F55 min		—
None req'd	F467M	Nuts for general use— metric	Naval brass UNS C46200	No	M6 - M36 mm	—	—	—	345 MPa	—	B65 min		—
None req'd	F467M	Nuts for general use— metric	Naval brass UNS C46400	No	M6 - M36 mm	—	—	—	345 MPa	—	B55 min		—
None req'd	F467M	Nuts for general use— metric	Phosphor bronze UNS C51000	No	M6 - M36 mm	—	—	—	415 MPa	—	B60 min		—
None req'd	F467M	Nuts for general use— metric	Aluminum bronze UNS C61400	No	M6 - M36 mm	—	—	—	520 MPa	—	B70 min		—
None req'd	F467M	Nuts for general use— metric	Aluminum bronze UNS C63000	No	M6 - M36 mm	—	—	—	690 MPa	—	B85 min		—

Grade ID mark	ASTM spec number	Fastener description	Material	Is mfgr's ID symbol req'd?	Nominal size range	Bolts, screws, studs			Nuts	Hardness		Remarks or footnotes
						Proof load	Yield strength (min)	Tensile strength (min)	Proof load	Brinell	Rockwell	
None req'd	F467M	Nuts for general use—metric	Aluminum silicon bronze UNS C64200	No	M6 - M36 mm	—	—	—	520 MPa	—	B75 min	—
None req'd	F467M	Nuts for general use—metric	Silicon bronze UNS C65100	No	M6 - M36 mm	—	—	—	485 MPa	—	B75 min	—
None req'd	F467M	Nuts for general use—metric	Silicon bronze UNS C65500	No	M6 - M36 mm	—	—	—	345 MPa	—	B60 min	—
None req'd	F467M	Nuts for general use—metric	Silicon bronze UNS C66100	No	M6 - M36 mm	—	—	—	485 MPa	—	B75 min	—
None req'd	F467M	Nuts for general use—metric	Manganese bronze UNS C67500	No	M6 - M36 mm	—	—	—	380 MPa	—	B60 min	—
None req'd	F467M	Nuts for general use—metric	Cupro-nickel UNS C71000	No	M6 - M36 mm	—	—	—	310 MPa	—	B50 min	—
None req'd	F467M	Nuts for general use—metric	Cupro-nickel UNS C71500	No	M6 - M36 mm	—	—	—	380 MPa	—	B60 min	—
None req'd	F467M	Nuts for general use—metric	Ni-Mo based UNS N10001	No	M6 - M36 mm	—	—	—	790 MPa	—	C20 min	—
None req'd	F467M	Nuts for general use—metric	Ni-Mo-Cr based UNS N10002	No	M6 - M36 mm	—	—	—	760 MPa	—	C20 min	—
None req'd	F467M	Nuts for general use—metric	Ni-Cu Class A UNS N04400	No	M6 - 36 mm	—	—	—	550 MPa	—	B75 min	—
None req'd	F467M	Nuts for general use—metric	Ni-Cu Class B UNS N04405	No	M6 - M36 mm	—	—	—	485 MPa	—	B60 min	—
None req'd	F467M	Nuts for general use—metric	Ni-Cu-Al based UNS N05500	No	M6 - M36 mm	—	—	—	900 MPa	—	C24 min	—
None req'd	F467M	Nuts for general use—metric	Aluminum 2024 UNS A92024	No	M6 - M36 mm	—	—	—	380 MPa	—	B70 min	—
None req'd	F467M	Nuts for general use—metric	Aluminum 6061 UNS A96061	No	M6 - M36 mm	—	—	—	275 MPa	—	B40 min	—
None req'd	F467M	Nuts for general use—metric	Aluminum 6262 UNS A96262	No	M6 - M36 mm	—	—	—	360 MPa	—	B60 min	—
None req'd	F467M	Nuts for general use—metric	Titanium Gr 1	No	M6 - M36 mm	—	—	—	275 MPa	—	Vickers 140 mm	—
None req'd	F467M	Nuts for general use—metric	Titanium Gr 2	No	M6 - M36 mm	—	—	—	380 MPa	—	Vickers 150 mm	—

Grade ID mark	ASTM spec number	Fastener description	Material	Is mfgr's ID symbol req'd?	Nominal size range (inch)	Bolts, screws, studs			Nuts	Hardness		Remarks or footnotes
						Proof load (psi)	Yield strength (min psi)	Tensile strength (min psi)	Proof load (psi)	Brinell	Rockwell	
None req'd	F467M	Nuts for general use—metric	Titanium Gr 4	No	M6 - M36 mm	—	—	—	590 MPa	—	Vickers 200 min	—
None req'd	F467M	Nuts for general use—metric	Titanium Gr 5	No	M6 - M36 mm	—	—	—	930 MPa	—	C30 min	—
None req'd	F467M	Nuts for general use—metric	Titanium Gr 7	No	M6 - M36 mm	—	—	—	380 MPa	—	Vickers 160 min	—
None req'd	F468	Bolts, hex cap screws, and studs for general use	ETP copper UNS C11000	No	¼ - 1½	—	10,000	30,000 min 50,000 max	—	—	F65/F90	—
None req'd	F468	Bolts, hex cap screws, and studs for general use	Brass UNS C27000	No	¼ - 1½	—	50,000	60,000 min 90,000 max	—	—	F55/F80	—
None req'd	F468	Bolts, hex cap screws, and studs for general use	Naval brass UNS C46200	No	¼ - 1½	—	25,000	50,000 min 80,000 max	—	—	B65/B90	—
None req'd	F468	Bolts, hex cap screws, and studs for general use	Naval brass UNS C46400	No	¼ - 1½	—	15,000	50,000 min 80,000 max	—	—	B55/B75	—
None req'd	F468	Bolts, hex cap screws, and studs for general use	Phosphor bronze UNS C51000	No	¼ - 1½	—	35,000	60,000 min 90,000 max	—	—	B60/B95	—
None req'd	F468	Bolts, hex cap screws, and studs for general use	Aluminum bronze UNS C61300	No	¼ - ½	—	50,000	80,000 min 110,000 max	—	—	B70/B95	—
					3/8 - 1½		45,000	75,000 min 105,000 max			B70/B95	
None req'd	F468	Bolts, hex cap screws, and studs for general use	Aluminum bronze UNS C61400	No	¼ - 1½	—	35,000	75,000 min 110,000 max	—	—	B70/B95	—

Grade ID mark	ASTM spec number	Fastener description	Material	Is mfgr's ID symbol req'd?	Nominal size range (inch)	Mechanical properties						Remarks or footnotes
						Bolts, screws, studs			Nuts	Hardness		
						Proof load (psi)	Yield strength (min psi)	Tensile strength (min psi)	Proof load (psi)	Brinell	Rockwell	
None req'd	F468	Bolts, hex cap screws, and studs for general use	Aluminum bronze UNS C63000	No	¼ - 1½	—	50,000	100,000 min 130,000 max	—	—	B85/B100	—
None req'd	F468	Bolts, hex cap screws, and studs for general use	Aluminum silicon bronze UNS C64200	No	¼ - 1½	—	35,000	75,000 min 110,000 max	—	—	B75/B95	—
None req'd	F468	Bolts, hex cap screws, and studs for general use	Silicon bronze UNS C65100	No	¼ - ¾	—	•55,000	70,000 min 100,000 max	—	—	B75/B95	—
					⅞ - 1½	—	40,000	55,000 min 90,000 max	—	—	B70/B95	—
None req'd	F468	Bolts, hex cap screws, and studs for general use	Silicon bronze UNS C65500	No	¼-1½	—	20,000	50,000 min 80,000 max	—	—	B60/B80	—
None req'd	F468	Bolts, hex cap screws, and studs for general use	Silicon bronze UNS C66100	No	¼ - 1½	—	35,000	70,000 min 100,000 max	—	—	B75/B95	—
None req'd	F468	Bolts, hex cap screws, and studs for general use	Manganese bronze UNS C67500	No	¼ - 1½	—	25,000	55,000 min 85,000 max	—	—	B60/B90	—
None req'd	F468	Bolts, hex cap screws, and studs for general use	Cupro-nickel UNS C71000	No	¼ - 1½	—	15,000	45,000 min 75,000 max	—	—	B50/B85	—
None req'd	F468	Bolts, hex cap screws, and studs for general use	Cupro-nickel UNS C71500	No	¼ - 1½	—	20,000	55,000 min 85,000 max	—	—	B60/B95	—
None req'd	F468	Bolts, hex cap screws, and studs for general use	Ni-Mo based UNS N10001	No	¼ - 1½	—	45,000	115,000 min 145,000 max	—	—	C20/C32	—

Grade ID mark	ASTM spec number	Fastener description	Is mfgr's ID symbol req'd?	Nominal size range (inch)	Bolts, screws, studs Proof load (psi)	Yield strength (min psi)	Tensile strength (min psi)	Nuts Proof load (psi)	Hardness Brinell	Rockwell	Remarks or footnotes
None req'd	F468	Bolts, hex cap screws, and studs for general use	Ni-Mo-Cr based UNS N10002	No	1/4 - 1 1/2	—	45,000	110,000 min 140,000 max	—	—	C20/C32	—
None req'd	F468	Bolts, hex cap screws, and studs for general use	Ni-Cu Class A UNS N04400	No	1/4 - 3/4	—	40,000	80,000 min 130,000 max	—	—	B75/C25	—
				7/8 - 1 1/2	—	30,000	70,000 min 130,000 max	—	—	B60/C25	
None req'd	F468	Bolts, hex cap screws, and studs for general use	Ni-Cu Class A UNS N04400 Hot formed product	No	1/4 - 1 1/2	—	30,000	70,000 min 120,000 max	—	—	B60/B95	—
None req'd	F468	Bolts, hex cap screws, and studs for general use	Ni-Cu Class B UNS N04405	No	1/4 - 1 1/2	—	30,000	70,000 min 125,000 max	—	—	B60/C20	—
None req'd	F468	Bolts, hex cap screws, and studs for general use	Ni-Cu-Al based UNS N05500	No	1/4 - 7/8	—	90,000	130,000 min 180,000 max	—	—	C24/C37	—
				1 - 1 1/2	—	85,000	130,000 min 180,000 max	—	—	C24/C37	
None req'd	F468	Bolts, hex cap screws, and studs for general use	Aluminum 2024 - UNS A92024	No	1/4 - 1 1/2	—	36,000	55,000 min 70,000 max	—	—	B70/B85	—
None req'd	F468	Bolts, hex cap screws, and studs for general use	Aluminum 6061 UNS A96061	No	1/4 - 1 1/2	—	31,000	37,000 min 52,000 max	—	—	B40/B50	—

Grade and material markings—Part IV

ASTM markings

The American Society for Testing and Materials, 1916 Race St. Philadelphia, PA 19103, sponsors development of specifications for fasteners used in general and special engineering applications. These specifications detail chemical and mechanical properties of material strength levels for fasteners and are generally specific in referencing the actual product covered. A full range of types of products of various styles, thread series, lengths, etc. can be produced to meet ASTM requirements and would be marked for grade and material identification as required.

ASTM Grade and material identification markings required by ASTM specifications

Grade ID mark	ASTM spec number	Fastener description	Material	Is mfgr's ID symbol req'd?	Nominal size range (inch)	Bolts, screws, studs Proof load (psi)	Yield strength (min psi)	Tensile strength psi	Nuts Proof load (psi)	Brinell	Rockwell	Remarks or footnotes
None req'd	F468	Bolts, hex cap screws, and studs for general use	Aluminum 7075 UNS A97075	No	¼ - 1½	—	50,000	61,000 min 76,000 max	—	—	B80/B90	—
None req'd	F468	Bolts, hex cap screws, and studs for general use	Titanium Gr 1	No	¼ - 1½	—	30,000	40,000 min 70,000 max	—	—	Vickers 140/160	—
None req'd	F468	Bolts, hex cap screws, and studs for general use	Titanium Gr 2	No	¼ - 1½	—	45,000	55,000 min 85,000 max	—	—	Vickers 160/180	—
None req'd	F468	Bolts, hex cap screws, and studs for general use	Titanium Gr 4	No	¼ - 1½	—	75,000	85,000 min 115,000 max	—	—	Vickers 200/220	—
None req'd	F468	Bolts, hex cap screws, and studs for general use	Titanium Gr 5	No	¼ - 1½	—	125,000	135,000 min 165,000 max	—	—	C30/C36	—
None req'd	F468	Bolts, hex cap screws, and studs for general use	Titanium Gr 7	No	¼ - 1½	—	45,000	55,000 min 85,000 max	—	—	Vickers 160/180	—
None req'd	F468M	Bolts, hex cap screws, and studs for general use (metric)	UT copper UNS C11000	No	M6 - M36 mm	—	70 MPa	205 min 345 max MPa	—	—	F65/F90	—

Grade ID mark	ASTM spec number	Fastener description	Is mfgr's ID symbol req'd?	Nominal size range (mm)	Bolts, screws, studs			Nuts	Hardness		Remarks or footnotes
					Proof load (MPa)	Yield strength (min MPa)	Tensile strength MPa	Proof load (MPa)	Brinell	Rockwell	
None req'd	F468M	Bolts, hex cap screws, and studs for general use —metric	No	M6 - M36	—	345	410 min 620 max	—	—	F55/F80	—
None req'd	F468M	Bolts, hex cap screws, and studs for general use —metric	No	M6 - M36	—	170	345 min 550 max	—	—	B65/B90	—
None req'd	F468M	Bolts, hex cap screws, and studs for general use —metric	No	M6 - M36	—	105	345 min 550 max	—	—	B55/B75	—
None req'd	F468M	Bolts, hex cap screws, and studs for general use —metric	No	M6 - M36	—	240	410 min 620 max	—	—	B60/B95	—
None req'd	F468M	Bolts, hex cap screws, and studs for general use —metric	No	M6 - M36	—	240	520 min 760 max	—	—	B70/B95	—
None req'd	F468M	Bolts, hex cap screws, and studs for general use —metric	No	M6 - M36	—	345	690 min 900 max	—	—	B85/B100	—
None req'd	F468M	Bolts, hex cap screws, and studs for general use —metric	No	M6 - M36	—	240	520 min 760 max	—	—	B75/B95	—
None req'd	F468M	Bolts, hex cap screws, and studs for general use —metric	No	M6 - M20 M24 - M36	—	380 275	480 min 690 max 380 min 620 max	—	—	B75/B95 B70/B95	—
None req'd	F468M	Bolts, hex cap screws, and studs for general use —metric	No	M6 - M36	—	140	345 min 550 max	—	—	B60/B80	—
None req'd	F468M	Bolts, hex cap screws, and studs for general use —metric	No	M6 - M36	—	240	480 min 690 max	—	—	B75/B95	—
None req'd	F468M	Bolts, hex cap screws, and studs for general use —metric	No	M6 - M36	—	170	380 min 590 max	—	—	B60/B90	—
None req'd	F468M	Bolts, hex cap screws, and studs for general use —metric	No	M6 - M36	—	105	310 min 520 max	—	—	B50/B85	—

Material column:
- Brass UNS C27000
- Naval brass UNS C46200
- Naval brass UNS C46400
- Phosphor bronze UNS C51000
- Aluminum bronze UNS C61400
- Aluminum bronze UNS C63000
- Aluminum silicon bronze UNS C64200
- Silicon bronze UNS C65100
- Silicon bronze UNS C65500
- Silicon bronze UNS C66100
- Manganese bronze UNS C67500
- Cupro-nickel UNS C71600

65

Grade ID mark	ASTM spec number	Fastener description	Material	Is mfgr's ID symbol req'd?	Nominal size range (mm)	Bolts, screws, studs			Nuts	Hardness		Remarks or footnotes
						Proof load (MPa)	Yield strength (min MPa)	Tensile strength MPa	Proof load	Brinell	Rockwell	
None req'd	F468M	Bolts, hex cap screws, and studs for general use —metric	Cupro nickel UNS C71500	No	M6 - M36	—	140	380 min 590 max	—	—	B60/B95	—
None req'd	F468M	Bolts, hex cap screws, and studs for general use —metric	Ni-Mo based UNS N10001	No	M6 - M36	—	310	790 min 1000 max	—	—	C20/C32	—
None req'd	F468M	Bolts, hex cap screws, and studs for general use —metric	Ni-Mo-Cr based UNS N10002	No	M6 - M36	—	310	760 min 970 max	—	—	C20/C32	—
None req'd	F468M	Bolts, hex cap screws, and studs for general use —metric	Ni-Cu Class A UNS N04400	No	M6 - M20 M24 - M36	—	275 205	550 min 900 max 480 min 900 max	—	—	B75/C25 B60/C25	—
None req'd	F468M	Bolts, hex cap screws, and studs for general use —metric	Ni-Cu Class A UNS N04400 Hot formed product	No	M6 - M36	—	205	480 min 830 max	—	—	B60/B95	—
None req'd	F468M	Bolts, hex cap screws, and studs for general use —metric	Ni-Cu Class B UNS N04405	No	M6 - M36	—	205	480 min 860 max	—	—	B60/C20	—
None req'd	F468M	Bolts, hex cap screws, and studs for general use —metric	Ni-Cu-Al based UNS N05500	No	M6 - M20 M24 - M36	—	620 590	900 min 1240 max 900 min 1240 max	—	—	C24/C37 C24/C37	—
None req'd	F468M	Bolts, hex cap screws, and studs for general use —metric	Aluminum 2024 UNS A92024	No	M6 - M36	—	250	380 min 480 max	—	—	B70/B85	—
None req'd	F468M	Bolts, hex cap screws, and studs for general use —metric	Aluminum 6061 UNS A96061	No	M6 - M36	—	215	260 min 360 max	—	—	B40/B50	—
None req'd	F468M	Bolts, hex cap screws, and studs for general use —metric	Aluminum 7075 UNS A97075	No	M6 - M36	—	345	420 min 520 max	—	—	B80/B90	—
None req'd	F468M	Bolts, hex head screws, and studs for general use —metric	Titanium Gr 1	No	M6 - M36	—	205	280 min 480 max	—	—	Vickers 140/160	—

Grade ID mark	ASTM spec number	Fastener description	Material	Is mfgr's ID symbol req'd?	Nominal size range (mm)	Mechanical properties						Remarks or footnotes
						Bolts, screws, studs			Nuts	Hardness		
						Proof load (MPa)	Yield strength (min MPa)	Tensile strength (min MPa)	Proof load	Brinell	Rockwell	
None req'd	F468M	Bolts, hex head screws, and studs for general use — metric	Titanium Gr 2	No	M6 - M36	—	310	380 min 590 max	—	—	Vickers 160/180	—
None req'd	F468M	Bolts, hex head screws, and studs for general use — metric	Titanium Gr 4	No	M6 - M36	—	520	590 min 790 max	—	—	Vickers 200/220	—
None req'd	F468M	Bolts, hex head screws, and studs for general use — metric	Titanium Gr 5	No	M6 - M36	—	860	930 min 1140 max	—	—	C30/C36	—
None req'd	F468M	Bolts, hex head screws, and studs for general use — metric	Titanium Gr 7	No	M6 - M36	—	310	380 min 590 max	—	—	Vickers 160/180	—
4.6	F568	Bolts, screws, studs for general engineering applications — metric	Low or medium carbon steel	Yes	M5 - M100	225	240	400	—	Vickers 120/220	B67/B95	K,LL,MM,NN
4.8	F568	Bolts, screws, studs for general engineering applications — metric	Low or medium carbon steel, partially or fully annealed as required	Yes	M16 - M16	310	340	420	—	Vickers 130/220	B71/B95	K,LL,MM,NN
5.8	F568	Bolts, screws, studs for general engineering applications — metric	Low or medium carbon steel, cold worked	Yes	M5 - M24	380	420	520	—	Vickers 160/220	B82/B95	K,LL,MM,NN
8.8	F568	Bolts, screws, studs for general engineering applications — metric	Medium carbon steel, quenched and tempered	Yes	M16 - M72	600	660	830	—	Vickers 255/336	C23/C34	K,LL,MM,NN
8.8	F568	Bolts, screws, studs for general engineering applications — metric	Low carbon martensite steel, quenched and tempered	Yes	M16 - M36	600	660	830	—	Vickers 255/336	C23/C34	K,LL,MM,NN
8.8.3	F568	Bolts, screws, studs for general engineering applications — metric	Atmospheric corrosion resistant steel, quenched and tempered	Yes	M16 - M36	600	660	830	—	Vickers 255/336	C23/C34	J,K,LL,MM,NN
9.8	F568	Bolts, screws (and studs M12 or larger) for general engineering applications — metric	Medium carbon steel, quenched and tempered	Yes	M16 - M16	650	720	900	—	Vickers 280/360	C27/C36	K,LL,MM,NN,OO
+ (OO)	F568	Studs for general engineering applications — metric	Medium carbon steel, quenched and tempered	Yes	less than M12	650	720	900	—	Vickers 280/360	C27/C36	K,LL,MM,NN,OO

Footnotes are grouped on the last page of this Part IV series

Grade ID mark	ASTM spec number	Fastener description	Material	Is mfgr's ID symbol req'd?	Nominal size range (mm)	Mechanical properties						Remarks or footnotes
						Bolts, screws, studs			Nuts	Hardness		
						Proof load (MPa)	Yield strength (min MPa)	Tensile strength (min MPa)	Proof load (MPa)	Brinell	Rockwell	
＋ (OO)	F568	Studs for general engineering applications - metric	Low carbon martensite steel, quenched and tempered	Yes	Less than M12	650	720	900	—	Vickers 280/360	C22/C36	K,LL,MM,NN,OO
9.8	F568	Bolts, screws (and studs M12 or larger) for general engineering applications—metric	Low carbon martensite steel, quenched and tempered	Yes	M16 - M16	650	720	900		Vickers 280/360	C22/C36	K,LL,MM,NN,OO
10.9	F568	Bolts, screws (and studs M12 or larger) for general engineering applications—metric	Medium carbon steel, quenched and tempered	Yes	M5 - M20	830	940	1040		Vickers 327/382	C33/C39	K,LL,MM,NN,OO
	F568		Medium carbon alloy steel, quenched and tempered	Yes	M5 - M100	830	940	1040		Vickers 327/382	C33/C39	K,LL,MM,NN,OO
□ (OO)	F568	Studs for general engineering applications—metric	Medium carbon or medium carbon alloy steel, quenched and tempered	Yes	Less than M12	830	940	1040		Vickers 327/382	C33/C39	K,LL,MM,NN,OO
□ (OO)	F568	Studs for general engineering applications—metric	Low carbon martensite steel, quenched and tempered	Yes	Less than M12	830	940	1040		Vickers 327/382	C33/C39	K,LL,MM,NN,OO
10.9	F568	Bolts, screws (and studs M12 or larger) for general engineering applications—metric	Low carbon martensite steel, quenched and tempered	Yes	M5 - M36	830	940	1040		Vickers 327/382	C33/C39	K,LL,MM,NN,OO
10.9.3	F568	Bolts, screws, studs for general engineering applications—metric	Atmospheric corrosion resistant steel, quenched and tempered	Yes	M16 - M36	830	940	1040		Vickers 327/382	C33/C39	J,LL,MM,NN,OO
12.9	F568	Bolts, screws (and studs M12 or larger) for general engineering applications—metric	Alloy steel, quenched and tempered	Yes	M16 - M100	970	1100	1220		Vickers 372/434	C38/C44	K,LL,MM,NN,OO
△ (OO)	F568	Studs for general engineering applications—metric	Alloy steel, quenched and tempered	Yes	Less than M12	970	1100	1220		Vickers 372/434	C38/C44	K,LL,MM,NN,OO

Footnotes are grouped on the last page of this Part IV series

Grade ID mark	ASTM spec number	Fastener description	Material	Is mfgr's ID symbol req'd?	Nominal size range (inch)	Mechanical properties						Remarks or footnotes
						Bolts, screws, studs			Nuts	Hardness		
						Proof load (psi)	Yield strength (min psi)	Tensile strength psi	Proof load (psi)	Brinell	Rockwell	
1 (PP)	F-593	Bolts, hex cap screws and studs	Stainless steel alloys 303, 303 Se, 304, 305, 384, XM1, XM7	No (PP)								1,PP
			• Cold worked		1/4 - 5/8	—	65,000	100,000/150,000	—	—	B95/C32	
					3/4 - 1 1/2	—	45,000	85,000/140,000	—	—	B80/C32	
			• Headed and rolled from annealed stock and then re-annealed		1/4 - 1 1/2	—	50,000 max (machined specimen)	85,000 max	—	—	B85 max	
			• Machined from annealed or solution annealed stock		1/4 - 1 1/2	—	30,000	75,000/100,000	—	—	B65/B95	
			• Machined from strain hardened stock		1/4 - 5/8	—	95,000	120,000/160,000	—	—	C24/C36	
					3/4 - 1	—	75,000	110,000/150,000	—	—	C20/C32	
					1 1/8 - 1 1/4	—	60,000	100,000/140,000	—	—	B95/C30	
					1 3/8 - 1 1/2	—	45,000	95,000/130,000	—	—	B90/C28	
2 (PP)	F-593	Bolts, hex cap screws and studs	Stainless steel alloy 316	No (PP)								1,PP
			• Cold worked		1/4 - 5/8	—	65,000	100,000/150,000	—	—	B95/C32	
					3/4 - 1 1/2	—	45,000	85,000/140,000	—	—	B80/C32	
			• Headed and rolled from annealed stock and then re-annealed		1/4 - 1 1/2	—	50,000 max (machined specimen)	85,000 max	—	—	B85 max	
			• Machined from annealed or solution annealed stock		1/4 - 1 1/2	—	30,000	75,000/100,000	—	—	B65/B95	
			• Machined from strain hardened stock		1/4 - 5/8	—	95,000	120,000/160,000	—	—	C24/C36	
					3/4 - 1	—	75,000	110,000/150,000	—	—	C20/C32	
					1 1/8 - 1 1/4	—	60,000	100,000/140,000	—	—	B95/C30	
					1 3/8 - 1 1/2	—	45,000	95,000/130,000	—	—	B90/C28	

Footnotes are grouped on the last page of this Part IV series

Grade ID mark	ASTM spec number	Fastener description	Material	Is mfgr's ID symbol req'd?	Nominal size range (inch)	Mechanical properties						Remarks or footnotes
						Bolts, screws, studs			Nuts	Hardness		
						Proof load (psi)	Yield strength (min psi)	Tensile strength psi	Proof load (psi)	Brinell	Rockwell	
3 (PP)	F593	Bolts, hex cap screws, and studs	Stainless steel alloys 321 and 347	No (PP)								1,PP
			• Cold worked		1/4 - 5/8	—	65,000	100,000/150,000	—	—	B95/C32	
					3/4 - 1 1/2	—	45,000	85,000/140,000	—	—	B80/C32	
			• Headed and rolled from annealed stock and then re-annealed		1/4 - 1 1/2	—	50,000 max (machined specimen)	85,000 max	—	—	B85 max	
			• Machined from annealed or solution annealed stock		1/4 - 1 1/2	—	30,000	75,000/100,000	—	—	B65/B95	
			• Machined from strain hardened stock		1/4 - 5/8	—	95,000	120,000/160,000	—	—	C24/C36	
					3/4 - 1	—	75,000	110,000/150,000	—	—	C20/C32	
					1 1/8 - 1 1/4	—	60,000	100,000/140,000	—	—	B95/C30	
					1 3/8 - 1 1/2	—	45,000	95,000/130,000	—	—	B90/C28	
4 (PP)	F593	Bolts, hex cap screws, and studs	Stainless steel, alloys 430 and 430F	No (PP)								1,PP
			• Machined from annealed or solution annealed stock		1/4 - 1 1/2	—	35,000	70,000/100,000	—	—	B65/B95	
5 (PP)	F593	Bolts, hex cap screws, and studs	Stainless steel, alloys 410, 416, and 416 Se	No (PP)								1,PP
			• Hardened and tempered at 1050 F min		1/4 - 1 1/2	—	90,000	110,000/140,000	—	—	C20/C30	
			• Hardened and tempered at 525 F min		1/4 - 1 1/2	—	120,000	160,000/190,000	—	—	C34/C45	
6 (PP)	F593	Bolts, hex cap screws, and studs	Stainless steel, alloy 431	No (PP)								1,PP
			• Hardened and tempered at 1050 F min		1/4 - 1 1/2	—	100,000	125,000/150,000	—	—	C25/C32	
			• Hardened and tempered at 525 F min		1/4 - 1 1/2	—	140,000	180,000/220,000	—	—	C40/C48	

Footnotes are grouped on the last page of this Part IV series

Grade ID mark	ASTM spec number	Fastener description	Material	Is mfgr's ID symbol req'd?	Nominal size range (inch)	Mechanical properties						Remarks or footnotes
						Bolts, screws, studs			Nuts	Hardness		
						Proof load (psi)	Yield strength (min psi)	Tensile strength psi	Proof load (psi)	Brinell	Rockwell	
7 (PP)	F593	Bolts, hex cap screws, and studs	Stainless steel, alloy 630 • Solution annealed and age hardened after forming	No (PP)	¼ - 1½	—	105,000	135,000 / 170,000	—	—	C28 / C38	I,PP
1 (PP)	F594	Nuts	Stainless steel, alloys 303, 303 Se, 304, 305, 384, XM1, and XM7	No (PP)								
			• Annealed after all threading		¼ - 1½	—	—	—	70,000 min	—	B85 max	I,PP
			• Machined from annealed or solution annealed stock		¼ - 1½	—	—	—	75,000 min	—	B65 / B95	
			• Cold worked		¼ - ⅝	—	—	—	100,000 min	—	B95 / C32	
					¾ - 1½	—	—	—	85,000 min	—	B80 / C32	
			• Machined from strain hardened stock		¼ - ⅝	—	—	—	120,000 min	—	C24 / C36	
					¾ - 1	—	—	—	110,000 min	—	C20 / C32	
					1⅛ - 1¼	—	—	—	100,000 min	—	B95 / C30	
					1⅜ - 1½	—	—	—	85,000 min	—	B90 / C28	
A	F541	Eyebolts	Alloy steel, forged, quenched, and tempered	Yes	¼ - 2½	Refer to complete F541 spec	70,000 min 100,000 max	95,000	—	197/248	B93 / B101	Markings are forged in raised characters.

ASTM footnotes

I. Excluding studs, all markings located on top of head raised or depressed

J. Manufacturer may add other distinguishing marks indicating the fastener is atmospheric corrosion resistant and of a weathering type

K. All markings shall be located on top of the head, raised or depressed. Base of property class symbols shall be positioned toward the closest periphery of the head

LL. Alternatively, for hex head products, markings may be indented on side of head with the base of the property class symbols positioned toward the bearing surface

MM. Bolts and screws of nominal thread diameters smaller than M5 need not be marked. Additionally, slotted and recessed screws of nominal thread diameters M5 and larger need not be marked. Metric bolts and screws shall not be marked with radial line symbols

NN. Studs of nominal thread diameters smaller than M5 need not be marked. Additionally, classes 4.6, 4.8, and 5.8 studs smaller than M12 need not be marked

OO. This is the grade mark symbol for studs of this property class in sizes M5 up to but not including M12

PP. Grade and manufacturer's identification symbols are required only when specified on the order

Grade and material markings—Part V

ASTM markings

The American Society for Testing and Materials, 1916 Race St. Philadelphia, PA 19103, sponsors development of specifications for fasteners used in general and special engineering applications. These specifications detail chemical and mechanical properties of material strength levels for fasteners and are generally specific in referencing the actual product covered. A full range of types of products of various styles, thread series, lengths, etc. can be produced to meet ASTM requirements and would be marked for grade and material identification as required.

ASTM Grade and material identification markings required by ASTM specifications

Grade ID mark	ASTM spec number	Fastener description	Material	Is mfgr's ID symbol req'd?	Nominal size range (inch)	Bolts, screws, studs			Nuts	Hardness		Remarks or footnotes
						Proof load (psi)	Yield strength (min psi)	Tensile strength psi	Proof load (psi)	Brinell	Rockwell	
2 (PP)	F-594	Nuts	Stainless steel, alloy 316	No (PP)								I,PP
			• Annealed after all threading		¼ - 1½	—	—	—	70,000 min	—	B85 / max	
			• Machined from annealed or solution annealed stock		¼ - 1½	—	—	—	75,000 min	—	B65 / B95	
			• Cold worked		¼ - ⅝	—	—	—	100,000 min	—	B95 / C32	
					¾ - 1½	—	—	—	85,000 min	—	B80 / C32	
			• Machined from strain hardened stock		¼ - ⅝	—	—	—	120,000 min	—	C24 / C36	
					¾ - 1	—	—	—	110,000 min	—	C20 / C32	
					1⅛ - 1¼	—	—	—	100,000 min	—	B95 / C30	
					1⅜ - 1½	—	—	—	85,000 min	—	B90 / C28	
3 (PP)	F-594	Nuts	Stainless steel, alloys 321, 347	No (PP)								I,PP
			• Annealed after all threading		¼ - 1½	—	—	—	70,000 min	—	B85 / max	
			• Machined from annealed or solution annealed stock		¼ - 1½	—	—	—	75,000 min	—	B65 / B95	
			• Cold worked		¼ - ⅝	—	—	—	100,000 min	—	B95 / C32	
					¾ - 1½	—	—	—	85,000 min	—	B80 / C32	

Footnotes are grouped on the last page of this Part V series

Grade ID mark	ASTM spec number	Fastener description	Material	Is mfgr's ID symbol req'd?	Nominal size range (inch)	Bolts, screws, studs Proof load (psi)	Yield strength (min)	Tensile strength (min)	Nuts Proof load (psi)	Brinell	Rockwell	Remarks or footnotes
			• Machined from strain hardened stock		¼ - ⅝	—	—	—	120,000 min	—	C24/C36	
					¾ - 1	—	—	—	110,000 min	—	C20/C32	
					1⅛ - 1¼	—	—	—	100,000 min	—	B95/C30	
					1⅜ - 1½	—	—	—	85,000 min	—	B90/C28	
4 (PP)	F594	Nuts	Stainless steel, alloys 430, 430F • Machined from annealed or solution annealed stock	No (PP)	¼ - 1½	—	—	—	70,000 min	—	B65/B95	I, PP
5 (PP)	F594	Nuts	Stainless steel alloys 410, 416, 416 Se • Hardened and tempered at 1050 F min	No (PP)	¼ - 1½	—	—	—	110,000 min	—	C20/C30	I, PP
			• Hardened and tempered at 525 F min		¼ - 1½	—	—	—	160,000 min	—	C34/C45	
6	F594	Nuts	Stainless steel, alloy 431 • Hardened and tempered at 1050 F min	No (PP)	¼ - 1½	—	—	—	125,000 min	—	C25/C32	I, PP
			• Hardened and tempered at 525 F min		¼ - 1½	—	—	—	180,000 min	—	C40/C48	
7	F594	Nuts	Stainless steel, alloy 630 • Solution annealed and age hardened after forming	No (PP)	¼ - 1½	—	—	—	135,000 min	—	C28/C38	I, PP
A1-50	F738	Bolts, screws, and studs —metric	Stainless steel, alloys 303, 303 Se, 304, 305, 384, XM1, XM7 • Headed and rolled from annealed stock and then re-annealed	Yes	M16 - M5 mm M6 - M36 mm	— —	— 210 MPa	500 MPa 500 MPa	— —	Vickers 155/220 Vickers 155/220	B81/B95 B81/B95	I, MM, QQ

Footnotes are grouped on the last page of this Part V series.

Grade ID mark	ASTM spec number	Fastener description	Material	Is mfgr's ID symbol req'd?	Bolts, screws, studs				Nuts	Hardness		Remarks or footnotes
					Nominal size range (mm)	Proof load (MPa)	Yield strength (Min MPa)	Tensile strength (min MPa)	Proof load (MPa)	Brinell	Rockwell	
A1-70	F738	Bolts, screws, and studs —metric	Stainless steel, alloys 303, 303 Se, 304, 305, 384, XM1, XM7 • Cold worked	Yes	M1.6 - M5	—	—	700	—	Vickers 220/330	B96/C33	I,MM,QQ
					M6 - M20	—	450	700	—	Vickers 220/330	B96/C33	
					Over M20 -M36	—	300	550	—	Vickers 160/310	B83/C31	
A1-80	F738	Bolts, screws, and studs —metric	Stainless steel, alloys 303, 303 Se, 304, 305, 384, XM1, XM7 • Machined from strain hardened stock	Yes	M1.6 - M5	—	—	800	—	Vickers 240/350	C23/C36	I,MM,QQ
					M6 - M20	—	600	800	—	Vickers 240/350	C23/C36	
					Over M20 - M24	—	500	700	—	Vickers 220/330	B96/C33	
					Over M24 - M30	—	400	650	—	Vickers 200/310	B93/C30	
					Over M30 - M36	—	300	600	—	Vickers 180/285	B89/C28	
A2-50	F738	Bolts, screws, and studs —metric	Stainless steel, alloys 321, 347 • Headed and rolled from annealed stock and then re-annealed	Yes	M1.6 - M5	—	—	500	—	Vickers 155/220	B81/B95	I,MM,QQ
					M6 - M36	—	210	500	—	Vickers 155/220	B81/B95	
A2-70	F738	Bolts, screws, and studs —metric	Stainless steel, alloys 321, 347 • Cold worked	Yes	M1.6 - M5	—	—	700	—	Vickers 220/330	B96/C33	I,MM,QQ
					M6 - M20	—	450	700	—	Vickers 220/330	B96/C33	
					Over M20 - M36	—	300	550	—	Vickers 160/310	B83/C31	
A2-80	F738	Bolts, screws, and studs —metric	Stainless steel, alloys 321, 347 • Machined from strain hardened stock	Yes	M1.6 - M5	—	—	800	—	Vickers 240/350	C23/C36	I,MM,QQ
					M6 - M20	—	600	800	—	Vickers 240/350	C23/C36	
					Over M20 - M24	—	500	700	—	Vickers 220/330	B96/C33	
					Over M24 - M30	—	400	650	—	Vickers 200/310	B93/C30	
					Over M30 - M36	—	300	600	—	Vickers 180/285	B89/C28	
A4-50	F738	Bolts, screws, and studs —metric	Stainless steel, alloy 316 • Headed and rolled from annealed stock and then re-annealed	Yes	M1.6 - M5	—	—	500	—	Vickers 155/220	B81/B95	I,MM,QQ
					M6 - M36	—	210	500	—	Vickers 155/220	B81/B95	

Footnotes are grouped on the last page of this Part V series.

Grade ID mark	ASTM spec number	Fastener description	Material	Is mfgr's ID symbol req'd?	Nominal size range (mm)	Bolts, screws, studs Proof load (MPa)	Yield strength (Min MPa)	Tensile strength (min MPa)	Nuts Proof load (MPa)	Hardness Brinell	Rockwell	Remarks or footnotes
A4-70	F738	Bolts, screws, and studs —metric	Stainless steel, alloy 316 • Cold worked	Yes	M1.6 - M5	—	—	700	—	Vickers 220/330	B96/C33	I,MM,QQ
					M6 - M20	—	450	700	—	Vickers 220/330	B96/C33	
					Over M20 - M36	—	300	550	—	Vickers 160/310	B83/C31	
A4-80	F738	Bolts, screws, and studs —metric	Stainless steel, alloy 316 • Machined from strain hardened stock	Yes	M1.6 - M5	—	—	800	—	Vickers 240/350	C23/C36	I,MM,QQ
					M6 - M20	—	600	800	—	Vickers 240/350	C23/C36	
					Over M20 - M24	—	500	700	—	Vickers 220/330	B96/C33	
					Over M24 - M30	—	400	650	—	Vickers 200/310	B93/C30	
					Over M30 - M36	—	300	600	—	Vickers 180/285	B89/C28	
F1-45	F738	Bolts, screws, and studs —metric	Stainless steel, alloys 430, 430F • Headed and rolled from annealed stock and then re-annealed	Yes	M1.6 - M5	—	—	450	—	Vickers 135/220	B74/B96	I,MM,QQ
					M6 - M36	—	250	450	—	Vickers 135/220	B74/B96	
F1-60	F738	Bolts, screws, and studs —metric	Stainless steel, alloys 430, 430F • Cold worked	Yes	M1.6 - M5	—	—	600	—	Vickers 180/285	B89/C28	I,MM,QQ
					M6 - M36	—	410	600	—	Vickers 180/285	B89/C28	
C1-50	F738	Bolts, screws, and studs —metric	Stainless steel, alloy 410 • Machined from annealed or solution annealed stock	Yes	M1.6 - M5	—	—	500	—	Vickers 155/220	B81/B96	I,MM,QQ
					M6 - M36	—	250	500	—	Vickers 155/220	B81/B96	
C1-70	F738	Bolts, screws, and studs —metric	Stainless steel, alloy 410 • Hardened and tempered at 565 C min	Yes	M1.6 - M5	—	—	700	—	Vickers 220/330	B96/C34	I,MM,QQ
					M6 - M36	—	410	700	—	Vickers 220/330	B96/C34	
C1-110	F738	Bolts, screws, and studs —metric	Stainless steel, alloy 410 • Hardened and tempered at 275 C min	Yes	M1.6 - M5	—	—	1100	—	Vickers 350/440	C36/C45	I,MM,QQ
					M6 - M36	—	820	1100	—	Vickers 350/440	C36/C45	

Footnotes are grouped on the last page of this Part V series

Grade ID mark	ASTM spec number	Fastener description	Material	Is mfgr's ID symbol req'd?	Nominal size range (mm)	Mechanical properties — Bolts, screws, studs — Proof load (MPa)	Yield strength (Min MPa)	Tensile strength (min MPa)	Nuts Proof load (MPa)	Hardness — Brinell/Vickers	Hardness — Rockwell	Remarks or footnotes
C3-80	F738	Bolts, screws, and studs - metric	Stainless steel, alloy 431 • Hardened and tempered at 565 C min	Yes	M1.6-M5 / M6-M36	— / —	— / 640	800 / 800	— / —	Vickers 240/340 / Vickers 240/340	C23/C35 / C23/C35	I,MM,QQ
C3-120	F738	Bolts, screws, and studs —metric	Stainless steel, alloy 431 • Hardened and tempered at 275 C min	Yes	M1.6-M5 / M6-M36	— / —	— / 950	1200 / 1200	— / —	Vickers 380/480 / Vickers 380/480	C39/C48 / C39/C48	I,MM,QQ
C4-50	F738	Bolts, screws, and studs —metric	Stainless steel, alloys 416, 416 Se • Machined from annealed or solution annealed stock	Yes	M1.6-M5 / M6-M36	— / —	— / 250	500 / 500	— / —	Vickers 155/220 / Vickers 155/220	B81/B96 / B81/B96	I,MM,QQ
C4-70	F738	Bolts, screws, and studs —metric	Stainless steel, alloys 416, 416 Se • Hardened and tempered at 565 C min	Yes	M1.6-M5 / M6-M36	— / —	— / 410	700 / 700	— / —	Vickers 220/330 / Vickers 220/330	B96/C34 / B96/C34	I,MM,QQ
C4-110	F738	Bolts, screws, and studs —metric	Stainless steel, alloys 416, 416 Se • Hardened and tempered at 275 C min	Yes	M1.6-M5 / M6-M36	— / —	— / 820	1100 / 1100	— / —	Vickers 350/440 / Vickers 350/440	C36/C45 / C36/C45	I,MM,QQ
P1-90	F738	Bolts, screws, and studs - metric	Stainless steel alloy 630 • Solution annealed and age hardened after forming	Yes	M1.6-M5 / M6-M36	— / —	— / 700	900 / 900	— / —	Vickers 285/370 / Vickers 285/370	C28/C38 / C28/C38	I,MM,QQ
None req'd	F835M	Hex socket head cap screws - metric	Alloy steel—quenched and tempered	No	M3-M20	—	1100	1040	—	Vickers 372/434	C38/C44	RR
A1-50	F836	Nuts—metric	Stainless steel—alloys 303, 303 Se, 304, 305, 384, XM1, XM7 • Machined from annealed or solution annealed stock, or formed and annealed	Yes	M1.6-M36				500	Vickers 155/220	B81/R95	SS

f ootnotes are grouped on the last page of this Part V series

Grade ID mark	ASTM spec number	Fastener description	Material	Is mfgr's ID symbol req'd?	Nominal size range (mm)	Mechanical properties						Remarks or footnotes
						Bolts, screws, studs			Nuts	Hardness		
						Proof load (MPa)	Yield strength (Min MPa)	Tensile strength (min MPa)	Proof load (MPa)	Brinell	Rockwell	
A1-70	F836	nuts – metric	Stainless steel— alloys 303, 303 Se, 304, 305, 384, XM1, XM7 • Cold worked	Yes	M16 - M20 / Over M20 - M36	— / —	— / —	— / —	700 / 550	Vickers 220/330 / Vickers 160/310	B96/C33 / B83/C31	SS
A1-80	F836	nuts – metric	Stainless steel alloys 303, 303 Se, 304, 305, 384, XM1, XM7 • Machined from strain hardened stock	Yes	M16 - M20 / Over M20 - M24 / Over M24 - M30 / Over 30 - M36	—	—	—	800 / 700 / 650 / 600	Vickers 240/350 / Vickers 220/330 / Vickers 200/310 / Vickers 180/285	C23/C36 / B96/C33 / B93/C30 / B89/C28	SS
A2-50	F836	nuts – metric	Stainless steel alloys 321, 347 • Machined from annealed or solution annealed stock, or formed and annealed	Yes	M16 - M36	—	—	—	500	Vickers 155/220	B81/B95	SS
A2-70	F836	nuts – metric	Stainless steel alloy 321, 347 • Cold worked	Yes	M16 - M20 / Over M20 - M36	—	—	—	700 / 550	Vickers 220/330 / Vickers 160/310	B96/C33 / B83/C31	SS
A2-80	F836	nuts – metric	Stainless steel, alloy 321, 347 • Machined from strain hardened stock	Yes	M16 - M20 / Over M20 - M24 / Over M24 - M30 / Over M30 - M36	—	—	—	800 / 700 / 650 / 600	Vickers 240/350 / Vickers 220/330 / Vickers 200/310 / Vickers 180/285	C23/C36 / B96/C33 / B93/C30 / B89/C28	SS
A4-50	F836	nuts – metric	Stainless steel, alloy 316 • Machined from annealed or solution annealed stock, or formed and annealed	Yes	M16 - M36	—	—	—	500	Vickers 155/220	B81/B95	SS

ASTM footnotes

LL Including studs, all markings located on top of thread. See ASTM A962 for both bottom of thread. Shall be tensile test.

MM Bolts and screws of nominal thread diameters shall (in that M3) need not be marked. Additionally, studs and threaded sections of nominal thread diameters M5 and larger need not be marked. Metric bolts and screws shall not be marked with radial line symbols.

PP Grade and manufacturer's identification symbols are required only which specified on the order.

QQ Identification marking of studs shall be on the stud within the point end.

SS Markings shall be on the top of nut, top of flange, or on one of the wrenching flats. Markings located on one of the wrenching flats shall be depressed. Markings on all other locations may be raised or depressed. Nuts in nominal thread diameters M1 and smaller need not be marked.

Grade and material markings—Part VI

ASTM markings

The American Society for Testing and Materials, 1916 Race St. Philadelphia, PA 19103, sponsors development of specifications for fasteners used in general and special engineering applications. These specifications detail chemical and mechanical properties of material strength levels for fasteners and are generally specific in referencing the actual product covered. A full range of types of products of various styles, thread series, lengths, etc. can be produced to meet ASTM requirements and would be marked for grade and material identification as required.

ASTM Grade and material identification markings required by ASTM specifications

Grade ID mark	ASTM spec number	Fastener description	Material	Is mfgr's ID symbol req'd?	Nominal size range (mm)	Bolts, screws, studs Proof load (MPa)	Yield strength (Min MPa)	Tensile strength (min MPa)	Nuts Proof load (MPa)	Hardness Brinell	Hardness Rockwell	Remarks or footnotes
A4-70	F836	Nuts—metric	Stainless steel, alloy 316, • Cold worked	Yes	M16 - M20 Over M20 - M36	---	---	---	700 550	Vickers 220/330 Vickers 160/310	B96/C33 B83/C31	SS
A4-80	F836	Nuts—metric	Stainless steel, alloy 316, • Machined from strain hardened stock	Yes	M16 - M20 Over M20 - M24 Over M24 - M30 Over M30 - M36	---	---	---	800 700 650 600	Vickers 240/350 Vickers 220/330 Vickers 200/310 Vickers 180/285	C23/C36 B96/C33 B93/C30 B89/C28	SS
F1-45	F836	Nuts—metric	Stainless steel, alloy 430, 430F • Machined from annealed or solution annealed stock, or formed and annealed	Yes	M1.6 - M36	---	---	---	450	Vickers 135/220	B74/B96	SS

Footnotes are grouped on the last page of this Part VI series

Grade ID mark	ASTM spec number	Fastener description	Material	Is mfgr's ID symbol req'd?	Nominal size range (mm)	Mechanical properties				Hardness		Remarks or footnotes
						Bolts, screws, studs			Nuts			
						Proof load (MPa)	Yield strength (Min MPa)	Tensile strength (min MPa)	Proof load (MPa)	Brinell	Rockwell	
C1-70	F836	Nuts—metric	Stainless steel, alloy 410 • Hardened and tempered at 565 C min	Yes	M16 - M36	—	—	—	700	Vickers 220/330	H96 / C34	SS
C1-110	F836	Nuts—metric	Stainless steel, alloy 410 • Hardened and tempered at 275 C min	Yes	M16 - M36	—	—	—	1100	Vickers 350/440	C36 / C45	SS
C3-80	F836	Nuts—metric	Stainless steel, alloy 431 • Hardened and tempered at 565 C min	Yes	M16 - M36	—	—	—	800	Vickers 240/340	C23 / C35	SS
C3-120	F836	Nuts—metric	Stainless steel, alloy 431 • Hardened and tempered at 275 C min	Yes	M16 - M36	—	—	—	1200	Vickers 380/480	C39 / C48	SS
C4-70	F836	Nuts—metric	Stainless steel, alloy 416, 416 Se • Hardened and tempered at 565 C min	Yes	M16 - M36	—	—	—	700	Vickers 220/330	H96 / C34	SS
C4-110	F836	Nuts—metric	Stainless steel, alloy 416, 416 Se • Hardened and tempered at 275 C min	Yes	M16 - M36	—	—	—	1100	Vickers 350/440	C36 / C45	SS

Footnotes are grouped on the last page of this Part VI series

Grade ID mark	ASTM spec number	Fastener description	Material	Is mfr's ID symbol req'd?	Nominal size range (mm)	Bolts, screws, studs			Nuts	Hardness		Remarks or footnotes
						Proof load (MPa)	Yield strength (Min MPa)	Tensile strength (min MPa)	Proof load (MPa)	Brinell	Rockwell	
P1-90	F836	Nuts metric	Stainless steel alloy 630 • Solution annealed and age hardened after forming	Yes	M16 - M36	—	—	—	900	Vickers 285/370	C28/C38	SS
A1-50	F837M	Socket head cap screws —metric	Stainless steel alloys 303, 304, 305, 384, XM1, XM7 • Annealed	Yes	M16 - M5, M6 - M36	—	— / 250	500 / 500	— / —	Vickers 155/220, Vickers 155/220	B70/B95, B70/B95	GG
A1-70	F837M	Socket head cap screws —metric	Stainless steel alloys 303, 304, 305, 384, XM1, XM7 • Cold worked	Yes	M16 - M5, M6 - M14, M16 - M36	—	— / 400 / 270	700 / 700 / 550	— / — / —	Vickers 220/330, Vickers 220/330, Vickers 160/310	B96/C33, B96/C33, B83/C30	GG
C1-110	F837M	Socket head cap screws —metric	Stainless steel alloy 410 • Heat treated	Yes	M16 - M5, M6 - M36	—	— / 820	1100 / 1100	— / —	Vickers 350/440, Vickers 350/440	C36/C45, C36/C45	GG
None req'd	F844	Plain (flat) washers for general use	Steel, unhardened	No	Thru 3"	—						—

ASTM Footnotes

GG. All screws with nominal diameters of 5 mm and larger require marking. Marking may be on the side or top of the head.

SS. Markings shall be on the top of nut, top of flange of one end of the studs, bearing flats. Markings specified on one of the wrenching flats shall be depressed. Markings on all other locations may be raised or depressed. Nuts in smaller thread diameters M16 and smaller need not be marked.

This concludes the ASTM grade marking compilation.

Grade and material markings—Part VII

SAE and GM markings

Several years ago the Society of Automotive Engineers, 400 Commonwealth Dr, Warrendale, PA 15096, developed a strength grading system for carbon and alloy steel commercial fasteners Today it is the most widely used and copied system in existence in this country. General requirements are presented in the following table. General Motors Corp issues standards which are broadly used outside this one company For this reason, GM cross references to SAE Grades are included in this listing.

SAE and GM Grade and material identification markings required by SAE and GM specifications

Grade ID mark	Spec number	Fastener description	Material	Is mfgr's ID symbol req'd?	Nominal size range (inch)	Mechanical properties						Remarks or footnote(s)
						Bolts, screws, studs			Nuts	Hardness		
						Proof load (psi)	Yield strength (min psi)	Tensile strength (min psi)	Proof load (psi)	Brinell	Rockwell	
None req'd	SAE J429 Grade 1 / GM 255-M	Bolts, screws, studs and U-bolts[A]	Low or medium carbon steel	Yes except studs	¼ - 1½	33,000[B]	36,000[C]	60,000	—	—	B70/B100	D Equivalent to ASTM A307, Grade A
None req'd	SAE J429 Grade 2 / GM 260-M	Bolts, screws, and studs	Low or medium carbon steel	Yes except studs	¼ - ¾	55,000[B]	57,000[C]	74,000	—	—	B80/B100	D
					Over ¾ - 1½	33,000	36,000[C]	60,000	—	—	B70/B100	
None req'd	SAE J429 Grade 4	Studs	Medium carbon cold drawn steel	No	¼ - 1½	65,000	100,000[C]	115,000	—	—	C22/C32	D
	SAE J429 Grade 5 / GM 280-M	Bolts, screws, and studs	Medium carbon steel, quenched and tempered	Yes except studs	¼ - 1	85,000	92,000[C]	120,000	—	—	C25/C34	D Equivalent to ASTM A449
					Over 1 - 1½	74,000	81,000[C]	105,000	—	—	C19/C30	
	SAE J429 Grade 5.1 (E)	Sems	Low or medium carbon steel, quenched and tempered	Yes	#6 - ⅝	85,000	—	120,000	—	—	C25/C40	D,F
	GM 275-M	Bolts and screws			#6 - ⅜						C23/C39	D
	SAE J429 Grade 5.2	Bolts and screws	Low carbon martensite steel, fully killed, fine grain, quenched and tempered	Yes	¼ - 1	85,000	92,000[C]	120,000	—	—	C26/C36	D

Footnotes are grouped on the last page of this Part VII series.

Footnotes are grouped on the last page of this Part VII series.

Grade ID mark	Spec number	Fastener description	Material	Is mfgr's ID symbol req'd?	Nominal size range (inch)	Bolts, screws, studs			Nuts	Hardness		Remarks or footnote(s)
						Proof load (psi)	Yield strength (min psi)	Tensile strength (min psi)	Proof load (psi)	Brinell	Rockwell	
(symbol)	SAE J429 Grade 7 / GM 290-M	Bolts and screws	Medium carbon alloy steel, quenched and tempered	Yes	¼ - 1½	105,000	115,000	133,000	--	--	C28/C34	Roll threaded after heat treatment. D,F
(symbol)	SAE J429 Grade 8 / GM 300-M	Bolts, screws, and studs	Medium carbon alloy steel, quenched and tempered	Yes except studs	¼ - 1½	120,000	130,000	150,000	--	--	C33/C39	Equivalent to ASTM A354, Grade BD. D
None req'd	SAE J429 Grade 8.1	Studs	Elevated temperature drawn steel medium carbon alloy of SAE 1541 (or 1541H steel)	No	¼ - 1½	120,000	130,000	150,000	--	--	--	D
(symbol)	SAE J429 Grade 8.2	Bolts and screws	Low carbon martensite steel, fully killed, fine grain, quenched and tempered	Yes	¼ - 1	120,000	130,000	150,000	--	--	--	D
—	GM 455 M	Bolts and screws	Corrosion resistant steel	Yes	¼ - 1½	40,000	--	55,000	--	143 min	B79 min	D
4.6	SAE J1199 / GM 500M (4.6)	Bolts, screws, studs and U bolts—metric	Low or medium carbon steel	Yes	M5 - M36 mm	225 MPa	240 MPa	400 MPa	--	--	B67/B87 and B86/B100	Approximately equivalent to SAE J429 Grade 1 and ASTM A307 Grade A. G
4.8	SAE J1199 / GM 500M (4.8)	Bolts, screws, sems and studs—metric	Low or medium carbon steel	Yes	M1.6 - M16 mm	310 MPa	--/340 MPa	420 MPa	--	--	B71/B87 and B71/B100	G
5.8	SAE J1199 / GM 500M (5.8)	Bolts, screws, and studs—metric	Low or medium carbon steel (cold worked)	Yes	M5 - M24 mm	380 MPa/420 MPa	--/420 MPa	520 MPa	--	--	B82/B95 and B82/B100	Approximately equivalent to SAE J429 Grade 2. G

Grade ID mark	Spec number	Fastener description	Material	Is mfgr's ID symbol req'd?	Nominal size range	Mechanical properties						Remarks or footnote(s)
						Bolts, screws, studs			Nuts	Hardness		
						Proof load	Yield strength (min)	Tensile strength (min)	Proof load	Brinell	Rockwell	
8.8	SAE J1199	Bolts, screws, and studs —metric	Medium carbon or medium carbon alloy steel, quenched and tempered	Yes	M16 - M36 mm	600 MPa	660° MPa	830 MPa	—	—	$\frac{C23}{C34}$	G
O (H)	GM 500M (8.8)	Studs—metric	Medium carbon or medium carbon alloy steel, quenched and tempered								$\frac{C24}{C34}$	Approximately equivalent to SAE J429 Grade 5 and ASTM A449
8.8		Bolts, screws, and studs —metric	Low carbon martensite steel, quenched and tempered									
9.8	SAE J1199	Bolts, screws, sems and studs —metric	Medium carbon steel, quenched and tempered	Yes	M16 - M16 mm	650 MPa	—	900 MPa	—	—	$\frac{C27}{C36}$	G
+ (H)		Studs —metric	Medium carbon steel, quenched and tempered				—					Approximately 9% stronger than SAE J429 Grade 5 and ASTM A449
9.8	GM 500M (9.8)	Bolts, screws, sems, and studs —metric	Low carbon martensite steel, quenched and tempered									
± (H)		Studs—metric	Low carbon martensite steel, quenched and tempered				420 MPa					
9|8 (i)		Same as sems, but no washers - metric	Medium carbon steel, quenched and tempered									
12.9	SAE J1199	Bolts, screws, and studs metric	Alloy steel, oil quenched and tempered	Yes	M16 - M36 mm	970 MPa	1100° MPa	1220 MPa	—	—	$\frac{C38}{C44}$	G
△ (H)												

Footnotes are grouped on the last page of this Part VII series.

83

Grade ID mark	Spec number	Fastener description	Material	Is mfgr's ID symbol req'd?	Nominal size range (inch)	Bolts, screws, studs — Proof load	Yield strength (min)	Tensile strength (min)	Nuts — Proof load (psi)	Hardness — Brinell	Hardness — Rockwell	Remarks or footnote(s)
10.9		Bolts, screws, and studs —metric	Medium carbon alloy steel, quenched and tempered		M6 - M36 mm							
□ (H)	SAE J1199	Studs—metric	Medium carbon alloy steel, quenched and tempered	Yes		830 MPa	940 MPa	1040 MPa	—	—	C33/C39	G Approximately equivalent to SAE J429, Grade 8 and ASTM A354, Grade BD
□ (H)	GM 500M (10.9)	Studs—metric	Low carbon martensite steel, quenched and tempered		M5 - M36 mm							
10.9		Bolts, screws, and studs —metric	Low carbon martensite steel, quenched and tempered									
None req'd	SAE J995 Grade 2 / GM 284M	Nuts	Low or medium carbon steel	No	¾ - 1½	—	—	—	90,000 / ---	—	C32 max / C30 max	
(nut, L,M)	SAE J995 Grade 5 / GM 286M	Nuts	Low or medium carbon steel	No	¾ - 1 / Over 1 - 1½	—	—	—	120,000 / 109,000 / 105,000 / 94,000	—	C32 max / C30 max	L,M
(nut, L,M)	SAE J995 Grade 8 / GM 301M	Nuts	Low or medium carbon steel	No	¾ - ⅞ / Over ⅝ - 1 / Over 1 - 1½ / Smaller than 1 / 1 and larger	—	—	—	120,000 / 109,000 / 115,000 / 104,000 / 105,000 / 94,000 / 150,000 / 150,000 / 150,000 / 150,000	—	C24/C32 / C26/C34 / C26/C36 / C24/C32 / C24/C34	L,M

Footnotes are grouped on the last page of this Part VII series.

Grade ID mark	Spec number	Fastener description	Material	Is mfr's ID symbol req'd?	Nominal size range (inch)	Mechanical properties						Remarks or footnote(s)
						Bolts, screws, studs			Nuts	Hardness		
						Proof load	Yield strength (min psi)	Tensile strength (psi)	Proof load (MPa)	Brinell	Rockwell	
(5)	GM 510M (5)	Nuts – metric	Non heat treated carbon steel	No	1.6 - 4 mm 5 - 6 mm 8 - 10 mm 12 - 16 mm 20 - 36 mm	— — — — —	— — — — —	— — — — —	520 580 590 610 630	—	B70 min / C30 max B78 min / C30 max	Coarse thread Style 1 hex nuts.
(9)	GM 510M (9)	Nuts – metric	Non heat treated carbon steel	No	3 - 4 mm 5 - 6 mm 8 - 10 mm 12 - 16 mm 20 - 36 mm	— — — — —	— — — — —	— — — — —	900 915 940 950 920	—	B85 min / C30 max B89 min / C30 max	Coarse thread Style 2 hex nuts
(10)	GM 510M (10)	Nuts – metric	Heat treated carbon steel	No	1.6 - 10 mm 12 - 16 mm 20 - 36 mm	— — —	— — —	— — —	1040 1050 1060	—	C26 min / C36 max	Coarse thread Style 1 hex nuts.
None req'd	SAE J430 Grade 0	Solid rivets	Carbon steel	No	All sizes	—	23,000	40,000-55,000	—	—	B65 max for sizes 1/16" and less.	
	SAE J430 Grade 1					—	28,000	52,000-62,000	—	—	B85 max for sizes 1/16" and less.	
	SAE J430 Grade 2					—	29,000	55,000-70,000	—	—	Not specified	
	SAE J430 Grade 3					—	38,000	68,000-82,000	—	—	Not specified	
None req'd	SAE J82 Grade 60M	Machine screws	Carbon steel	No	#4 - 1/4	—	—	60,000 min	—	—	B70 / B100	
None req'd	SAE J82 Grade 120M	Machine screws	Carbon steel, quenched and tempered	No	#4 - 1/4	—	—	120,000 min	—	—	C25 / C38	

SAE & GM footnotes

A. Whenever the yield strength appears, that is also implied.

B. Requirements for proof load apply only to stress relieved products.

C. Value applies to machine head specimens.

D. Unslotted bolts, screws, and hex head sems shall be grade marked as required. In addition, bolts and hex head sems shall be marked with the manufacturer's identification symbol. Marking shall be located on top of the head, and may be either raised or depressed at the manufacturer's option. A slotted hex head sems need not be marked.

E. Sems and similar products without washers.

F. Hex washer head and hex flange products without are standard washers shall have core hardness not exceeding Rockwell C38 and surface hardness not exceeding Rockwell C43.

G. Slotted and most recessed screws of all sizes and all other screws and bolts of sizes smaller than M5 need not be marked. All other bolts and screws of sizes M5 and larger shall be marked to identify property class and manufacturer. Markings shall be located on top of the head, and shall, at the option of the manufacturer, be either raised or depressed.

H. This is the optional property class symbol for studs of this property class in sizes M5 through M11.

I. Products made same as sems except without washers, shall be additionally identified with, at most of 12-headed fasteners, the property class marking as below: 42

J. Values apply to 1/16" and M10 thread series.

K. Values apply to 1/16" 1/2-1.0 and finer thread series.

L. All grades of hex jam, heavy hex jam, hex slotted, heavy hex slotted, hex thick and heavy hex nuts are not required to be grade marked.

M. Grade markings for nuts, fabricated by cutting from bar, tool shall consist of notches at the hexagon corners, and notches at each corner for Grade 5, and two notches at each corner for Grade 8.

Grade and material markings—Part VIII

ISO markings

ISO (the International Organization for Standardization) is a federation of the national standards bodies of the countries of the world. Purpose of developing international standards is to form the basis of a one-world system of engineering practices. It is intended that international decisions being documented as ISO standards will become accepted into the national standards of ISO member countries. Copies of ISO standards are available from American National Standards Institute (ANSI), 1430 Broadway, New York, NY 10018

ISO Identification markings required by ISO standards for externally threaded fasteners

Property class ID mark	ISO standard number[A]	Fastener description	Is mfgr's ID symbol req'd?	Nominal size range (mm)	Mechanical requirements						Remarks or footnotes
					Externally threaded fasteners			Rockwell hardness			
					Proof load stress MPa	Yield strength (min MPa)	Tensile strength (min MPa)	Surface (Max)	Core (Min)	Core (Max)	
4.6	ISO 898/1	Bolts, screws and studs	Low or medium carbon steel	Yes	M5-M100	225	240	400	---	B67	B95	B,C
4.8	ISO 898/1	Bolts, screws and studs	Low or medium carbon steel, fully or partially annealed	Yes	M1.6-M16	310	340	420	---	B71	B95	B,C
5.8	ISO 898/1	Bolts, screws and studs	Low or medium carbon steel, cold worked	Yes	M5-M24	380	420	520	---	B82	B95	B,C,D
8.8	ISO 898/1	Bolts, screws and studs	Medium carbon steel, quenched and tempered	Yes	M16-M72	600	660	830	30N56	C23	C34	B,C
8.8 (underlined)	ISO 898/1	Bolts, screws and studs	Low carbon martensite steel, quenched and tempered	Yes	M16-M36	600	660	830	30N56	C23	C34	B,C
9.8	ISO 898/1	Bolts, screws (and studs M12 or larger)	Medium carbon steel, quenched and tempered	Yes	M1.6-M16	650	720	900	30N58	C27	C36	B,C

Footnotes are grouped on the last page of this Part VIII series.

Property class ID mark	ISO standard number[A]	Fastener description	Material	Is mfgr's ID symbol req'd?	Nominal size range (mm)	Mechanical requirements						Remarks or footnotes
						Externally threaded fasteners			Rockwell hardness			
						Proof load stress MPa	Yield strength (min MPa)	Tensile strength (min MPa)	Surface (Max)	Core (Min)	Core (Max)	
+	ISO 898/1	Studs of class 9.8	Medium carbon steel, quenched and tempered	Yes	Less than M12	650	720	900	30N58	C27	C36	C,E
9.8	ISO 898/1	Bolts, screws (and studs M12 or larger)	Low carbon martensite steel, quenched and tempered	Yes	M16-M16	650	720	900	30N58	C27	C36	B,C
±	ISO 898/1	Studs of class 9.8	Low carbon martensite steel, quenched and tempered	Yes	Less than M12	650	720	900	30N58	C27	C36	C,E
10.9	ISO 898/1	Bolts, screws (and studs M12 or larger)	Medium carbon steel, quenched and tempered	Yes	M5-M20	830	940	1040	30N59	C33	C39	B,C
□	ISO 898/1		Medium carbon alloy steel, quenched and tempered	Yes	M5-M100	830	940	1040	30N59	C33	C39	B,C
□	ISO 898/1	Studs of class 10.9	Medium carbon or medium carbon alloy steel, quenched and tempered	Yes	Less than M12	830	940	1040	30N59	C33	C39	C,E
10.9	ISO 898/1	Bolts, screws (and studs M12 or larger)	Low carbon martensite steel, quenched and tempered	Yes	M5-M36	830	940	1040	30N59	C33	C39	B,C
□	ISO 898/1	Studs of class 10.9	Low carbon martensite steel, quenched and tempered	Yes	Less than M12	830	940	1040	30N59	C33	C39	C,E
12.9	ISO 898/1	Bolts, screws (and studs M12 or larger)	Alloy steel, quenched and tempered	Yes	M16-M100	970	1100	1220	30N63	C38	C44	B,C,F
△	ISO 898/1	Studs of class 12.9	Alloy steel, quenched and tempered	Yes	Less than M12	970	1100	1220	30N63	C38	C44	C,E

Footnotes are grouped on the last page of this Part VIII series.

ISO

Identification markings required by ISO standards for internally threaded fasteners.

Property class ID mark	Property class of nut[G]	Dimensional style of nut	Material	Is mfgr's ID symbol req'd?	Nominal size range (mm)	Proof load stress (MPa) Non overtapped nuts	Proof load stress (MPa) Overtapped nuts	Rockwell hardness (Min)	Rockwell hardness (Max)	Remarks or footnotes
None req'd	04	Hex jam	Carbon steel	No	M5-M36	380		B89	C30	
05	05	Hex jam	Carbon steel	Yes	M5-M36	500	—	C26	C36	H
5	5	Hex, Style 1	Carbon steel	Yes	M1.6-M4	520		B70	C30	H
					M5, M6	580	465			
					M8, M10	590	470			
					M12-M16	610	490			
		Heavy hex			M20-M36	630	500	B78	C30	
					M42-M100	630	500	B70	C30	
9	9	Hex, Style 2	Carbon steel	Yes	M3-M4	900	—	B85	C30	H,I
		Hex, Style 2 and hex flange			M5, M6	915	—			
					M8, M10	940	—	B89	C30	
					M12-M16	950	—			
		Hex, Style 2			M20	920	—			
		Heavy hex			M24-M36	920	—			
					M42-M100	920	—			
10	10	Hex, Style 1	Alloy steel quenched and tempered	Yes	M1.6-M4	1040	—	C26	C36	H,I
					M5-M10	1040	—			
		Hex, Style 1 and hex flange			M12-M16	1050	—			
					M20	1060	—			
		Hex, Style 1			M24-M36	1060	—			
12	12	Hex, Style 2	Alloy steel quenched and tempered	Yes	M3-M4	1150	—	C26	C36	H,I
					M5, M6	1150	920			
					M8, M10	1160	930			
					M12-M16	1190	950			
					M20-M36	1200	960			
		Heavy hex			M42-M100	1200	960			

Footnotes are on next page

ISO footnotes from preceding tables

A. Although ISO 898/1 presents 10 property classes. IFI has been unable to identify any commercial or industrial need by North American industry for ISO property classes 3.6, 5.6 and 6.8 bolts, screws and studs.

B. Marking is required for hex bolts and screws with nominal diameters $\geq$5mm where shape of fastener allows marking to be accomplished, preferably on the head; alternatively on the side of head by indenting

C. Marking is required for studs with nominal diameters equal to or greater than 5mm, preferably on the extreme end of the threaded portion by indenting. For studs with interference fit, the marking shall be at the nut end.

D. Class 5.8 applies only to bolts and screws with lengths 150mm and shorter and to studs of all lengths.

E. This is the grade mark symbol for studs of this property class in sizes M5 up to but not including M12

F. Caution is advised when considering use of Class 12.9 products. Capability of the fastener manufacturer, as well as anticipated service environment, should be carefully considered. Class 12.9 products require rigid control of heat treating operations and careful monitoring of as-quenched hardness, surface discontinuities, depth of partial decarburization, and freedom from carburization. Stress corrosion cracking susceptibility also needs to be addressed.

G. All data was extracted from ASTM A563M and ISO 898/2. All values are as given in A563M. Values for property classes 04, 05, 5, 9, 10 and 12 non-overtapped nuts in sizes M36 and smaller are in both documents and are identical. Other classes, sizes and overtapped nut values are unique to A563M.

H. Hex nuts of thread diameters >5mm and property classes equal to or higher than 8 and property class 05 shall be marked as noted, by indenting on the side or bearing surface, or by embossing on the chamfer

I. Alternative marking system according to clock-face system is as follows:

Property class	9	10	12
Alternative marking — either designation symbol	9	10	12
Alternative marking — or code symbol (clock-face system)			

Appendix B
Bolt Ultimate Shear and Tensile Strengths

[From ref. 18]

TABLE 8.1.5(a). *Ultimate Single Shear Strength of Threaded Steel Fasteners*

| Fastener diameter | | Basic shank area | Shear strength of fastener, ksi — Ultimate single shear strength, lbs.[a] | | | | | | | | | |
In.	b		35	38	75	91	95	108	125	132	156	180
0.112	# 4	0.0098520	345	374	739	897	936	1 064	1 232	1 300	1 537	1 773
0.125	1/8	0.012272	430	466	920	1 117	1 166	1 325	1 534	1 620	1 914	2 209
0.138	# 6	0.014957	523	568	1 122	1 361	1 421	1 615	1 870	1 974	2 333	2 692
0.156	5/32	0.019175	671	729	1 438	1 745	1 822	2 071	2 397	2 531	2 991	3 452
0.164	# 8	0.021124	739	803	1 584	1 922	2 007	2 281	2 640	2 788	3 295	3 802
0.188	3/16	0.027612	966	1 049	2 071	2 513	2 623	2 982	3 452	3 645	4 310	4 970
0.190	#10	0.028353	992	1 077	2 126	2 580	2 694	3 062	3 544	3 743	4 420	5 100
0.216	#12	0.036644	1 283	1 392	2 748	3 335	3 481	3 958	4 580	4 840	5 720	6 600
0.219	7/32	0.037582	1 315	1 428	2 819	3 420	3 570	4 060	4 700	4 960	5 860	6 760
0.250	1/4	0.049087	1 718	1 865	3 682	4 470	4 660	5 300	6 140	6 480	7 660	8 840
0.312	5/16	0.076699	2 684	2 915	5 750	6 980	7 290	8 280	9 590	10 120	11 970	13 810
0.375	3/8	0.11045	3 866	4 200	8 280	10 050	10 490	11 930	13 810	14 580	17 230	19 880
0.438	7/16	0.15033	5 260	5 710	11 270	13 680	14 280	16 240	18 790	19 840	23 450	27 060
0.500	1/2	0.19635	6 870	7 460	14 730	17 870	18 650	21 210	24 540	25 920	30 630	35 340
0.562	9/16	0.24850	8 700	9 440	18 640	22 610	23 610	26 840	31 060	32 800	38 770	44 700
0.625	5/8	0.30680	10 740	11 660	23 010	27 920	29 150	33 130	38 350	40 500	47 900	55 200
0.750	3/4	0.44179	15 460	16 790	33 130	40 200	42 000	47 700	55 200	58 300	68 900	79 500
0.875	7/8	0.60132	21 050	22 850	45 100	54 700	57 100	64 900	75 200	79 400	93 800	108 200
1.000	1	0.78540	27 490	29 850	58 900	71 500	74 600	84 800	98 200	103 700	122 500	141 400
1.125	1-1/8	0.99402	34 790	37 770	74 600	90 500	94 400	107 400	124 300	131 200	155 100	178 900
1.250	1-1/4	1.2272	43 000	46 600	92 000	111 700	116 600	132 500	153 400	162 000	191 400	220 900
1.375	1-3/8	1.4849	52 000	56 400	111 400	135 100	141 100	160 400	185 600	196 000	231 600	267 300
1.500	1-1/2	1.7671	61 800	67 100	132 500	160 800	167 900	190 800	220 900	233 300	275 700	318 100

[a]Values with the first digit <4 are shown to 4 significant figures, all other are shown to 3 significant figures.

[b]Fractional equivalent or screw number.

TABLE 8.1.5(b₁). *Ultimate Tensile Strength of Threaded Steel Fasteners*

Fastener diameter		Nominal minor area*	Ultimate tensile strength, lbs.[a,b,c]						
In.	d		55	62	62.5	125	140	160	180
						MIL-S-7742			
0.112	4-40	0.0050896	280	316	318	636	713	814	916
0.138	6-32	0.0076821	423	476	480	960	1 075	1 229	1 383
0.164	8-32	0.012233	673	758	765	1 529	1 713	1 957	2 202
0.190	10-32	0.018074	994	1,121	1 130	2 259	2 530	2 892	3 253
0.250	1/4-28	0.033394	1 837	2 070	2 087	4 170	4 680	5 340	6 010
0.312	5/16-24	0.053666	2 952	3 327	3 354	6 710	7 510	8 590	9 660
0.375	3/8-24	0.082397	4 530	5 110	5 150	10 300	11 540	13 180	14 830
0.438	7/16-20	0.11115	6 110	6 890	6 950	13 890	15 560	17 780	20 010
0.500	1/2-20	0.15116	8 310	9 370	9 450	18 900	21 160	24 190	27 210
0.562	9/16-18	0.19190	10 550	11 900	11 990	23 990	26 870	30 700	34 540
0.625	5/8-18	0.24349	13 390	15 100	15 220	30 440	34 090	38 960	43 800
0.750	3/4-16	0.35605	19 580	22 080	22 250	44 500	49 800	57 000	64 100
0.875	7/8-14	0.48695	26 780	30 190	30 430	60 900	68 200	77 900	87 700
1.000	1-12	0.63307	34 820	39 250	39 570	79 100	88 600	101 300	114 000
1.125	1-1/8-12	0.82162	45 200	50 900	51 400	102 700	115 000	131 500	147 900
1.250	1-1/4-12	1.0347	56 900	64 200	64 700	129 300	144 900	165 600	186 200
1.375	1-3/8-12	1.2724	70 000	78 900	79 500	159 000	178 100	203 600	229 000
1.500	1-1/2-12	1.5345	84 400	95 100	95 900	191 800	214 800	245 500	276 200

aValues shown above heavy line are for 2A threads; all other values are for 3A threads.

bNuts designed to develop the ultimate tensile strength of the fastener are required to develop the tabulated tension loads.

cValues with the first digit <4 are shown to 4 significant figures; all others are shown to 3 significant figures.

dFractional equivalent or number and threads per inch.

*Area computed using nominal minor diameter as published in Table 2.2.1 of Handbook H-28.

TABLE 8.1.5(b₂). *Ultimate Tensile Strength of Threaded Steel Fasteners (Continued)*

Fastener diameter		Maximum minor area[e]	Ultimate tensile strength, lbs [a,b,c]				
In.	d		Tensile strength of fastener, ksi				
			160	180	220	260	300
					MIL-S-8879		
0.112	4-40	0.0054367	869	979	1,196	1,414	1,631
0.138	6-32	0.0081553	1,305	1,468	1,794	2,120	2,447
0.164	8-32	0.012848	2,055	2,313	2,827	3,340	3,854
0.190	10-32	0.018602	2,976	3,348	4,090	4,840	5,580
0.250	1/4-28	0.034241	5,480	6,160	7,530	8,900	10,270
0.312	5/16-24	0.054905	8,780	9,880	12,080	14,280	16,470
0.375	3/8-24	0.083879	13,420	15,100	18,450	21,810	25,160
0.438	7/16-20	0.11323	18,120	20,380	24,910	29,440	33,970
0.500	1/2-20	0.15358	24,570	27,640	33,790	39,930	46,100
0.562	9/16-18	0.19502	31,200	35,100	42,900	50,700	58,500
0.625	5/8-18	0.24700	39,520	44,500	54,300	64,200	74,100
0.750	3/4-16	0.36082	57,700	64,900	79,400	93,800	108,200
0.875	7/8-14	0.49327	78,900	88,800	108,500	128,300	148,000
1.000	1-12	0.64156	102,600	115,500	141,100	166,800	192,500
1.125	1-1/8-12	0.83129	133,000	149,600	182,900	216,100	249,400
1.250	1-1/4-12	1.0456	167,300	188,200	230,000	271,900	313,700
1.375	1-3/8-12	1.2844	205,500	231,200	282,600	333,900	385,300
1.500	1-1/2-12	1.5477	247,600	278,600	340,500	402,400	464,300

[a] All values are for 3A threads.

[b] Nuts designed to develop the ultimate tensile strength of the fastener are required to develop the tabulated tension loads.

[c] Values with the first digit <4 are shown to 4 significant figures; all others are shown to 3 significant figures.

[d] Fractional equivalent or number and threads per inch.

[e] Area computed using maximum minor diameter as published in Tables II and III of MIL-S-8879.

Appendix C
Blind Rivet Requirements

BLIND RIVETS SHALL BE USED IN COMPLIANCE WITH THE JOINT ALLOWABLE TABLES IN MIL-HDBK-5, CHAPTER 8.

BLIND RIVETS SHALL CONFORM TO THE FOLLOWING REQUIREMENTS:

1. THE HOLE SIZE FOR BLIND INSTALLATION SHALL BE WITHIN THE LIMITS SPECIFIED ON THE APPLICABLE SPECIFICATION SHEET, STANDARD, OR DRAWING.

2. FOR DIMPLED ASSEMBLY, THE RIVET HOLES SHALL BE SIZED AFTER THE SHEETS HAVE BEEN DIMPLED.

3. MECHANICALLY LOCKED SPINDLE BLIND RIVETS (LOCKING RING OR COLLAR) MAY BE USED ON AIRCRAFT IN AIR INTAKE AREAS AND IN THE AREA FORWARD OF THE ENGINE.

4. FOR REPAIR AND REWORK, THE BLIND RIVETS USED IN REPLACEMENT OF SOLID SHANK RIVETS SHALL BE OVERSIZE OR ONE STANDARD SIZE LARGER (SEE REQMT 5).

5. OVERSIZE BLIND RIVETS MAY BE USED FOR REPAIR AND REWORK:

   a. OVERSIZE RIVETS ARE FOR USE IN NON-STANDARD HOLE DIAMETERS. NON-STANDARD HOLES ARE THE RESULT OF HOLE RESIZING DURING REWORK OR REPAIR, OR DUE TO MANUFACTURING ERROR IN NEW DESIGN.

   b. THE GRIP LENGTH OF THE OVERSIZE RIVET, THE BACKSIDE CLEARANCE (INSTALLED AND UNINSTALLED), AND THE PERFORMANCE CHARACTERISTICS SHALL BE EQUAL TO THE STANDARD RIVET THAT IS BEING REPLACED.

6. BLIND RIVETS SHALL NOT BE USED:

   a. IN FLUID TIGHT AREAS.

   b. ON AIRCRAFT CONTROL SURFACE HINGES, HINGE BRACKETS, FLIGHT CONTROL ACTUATING SYSTEMS, WING ATTACHMENT FITTINGS, LANDING GEAR FITTINGS, OR OTHER HEAVILY STRESSED LOCATIONS ON THE AIRCRAFT.

7. FRICTION LOCKED BLIND RIVETS (NO LOCKING RING OR COLLAR) SHALL NOT BE USED ON AIRCRAFT IN AIR INTAKE AREAS WHERE RIVET PARTS MAY BE INGESTED BY THE ENGINE.

8. NICKEL-COPPER ALLOY (MONEL) RIVETS WITH CADMIUM PLATING SHALL NOT BE USED WHERE THE AMBIENT TEMPERATURE IS ABOVE 400°F.

9. FLUSH HEAD RIVETS SHALL NOT BE MILLED TO OBTAIN FLUSHNESS WITH THE SURROUNDING SHEET WITHOUT PRIOR WRITTEN APPROVAL FROM THE DESIGN ACTIVITY.

10. OVERSIZE BLIND RIVETS SHALL NOT BE SPECIFIED IN NEW DESIGN. AN OVERSIZE BLIND RIVET IS ONE SPECIFICALLY DESIGNED FOR REPLACEMENT PURPOSES. ITS SHANK DIAMETER DIMENSION IS GREATER THAN A STANDARD BLIND RIVET.

11. CHEMICALLY EXPANDED BLIND RIVETS SHALL NOT BE USED.

THIS IS A DESIGN STANDARD, NOT TO BE USED AS A PART NUMBER.

(C) REWRITTEN

APPROVED 11 APR 52   REVISED (C) 19 OCT R4

P.A. NAVY — AS Other Cust	TITLE	MILITARY STANDARD
ARMY - AV USAF - 11	RIVETS, BLIND, STRUCTURAL, MECHANICALLY LOCKED AND FRICTION RETAINER SPINDLE, (RELIABILITY AND MAINTAINABILITY) DESIGN AND CONSTRUCTION REQUIREMENTS FOR.	MS33522
PROCUREMENT SPECIFICATION	SUPERSEDES:	SHEET 1 OF 1

NASA National Aeronautics and Space Administration	Report Documentation Page

1. Report No. NASA RP-1228	2. Government Accession No.	3. Recipient's Catalog No.
4. Title and Subtitle Fastener Design Manual		5. Report Date March 1990
		6. Performing Organization Code
7. Author(s) Richard T. Barrett		8. Performing Organization Report No. E-4911
		10. Work Unit No.
9. Performing Organization Name and Address National Aeronautics and Space Administration Lewis Research Center Cleveland, Ohio 44135-3191		11. Contract or Grant No.
		13. Type of Report and Period Covered Reference Publication
12. Sponsoring Agency Name and Address National Aeronautics and Space Administration Washington, D.C. 20546-0001		14. Sponsoring Agency Code

15. Supplementary Notes

16. Abstract

This manual was written for design engineers to enable them to choose appropriate fasteners for their designs. Subject matter includes fastener material selection, platings, lubricants, corrosion, locking methods, washers, inserts, thread types and classes, fatigue loading, and fastener torque. A section on design criteria covers the derivation of torque formulas, loads on a fastener group, combining simultaneous shear and tension loads, pullout load for tapped holes, grip length, head styles, and fastener strengths. The second half of this manual presents general guidelines and selection criteria for rivets and lockbolts.

17. Key Words (Suggested by Author(s)) Fastener design; Washers; Inserts; Torque table; Rivets; Lockbolts	18. Distribution Statement Unclassified – Unlimited Subject Category 37

19. Security Classif. (of this report) Unclassified	20. Security Classif. (of this page) Unclassified	21. No of pages 100	22. Price* A05

NASA FORM 1626 OCT 86   *For sale by the National Technical Information Service, Springfield, Virginia 22161

NASA-Langley, 1990

www.ingramcontent.com/pod-product-compliance
Lightning Source LLC
Chambersburg PA
CBHW081509170526
45166CB00008B/2601